S0-BWY-743

MATHEMATICAL TABLES & FORMULAE

Mathematical Tables and Formulae

BY F. J. CAMM

PHILOSOPHICAL LIBRARY

New York

15 EAST 40TH STREET, NEW YORK 16, N. Y.

Printed in the United States of America

PREFACE

This book has been specially compiled for the use of students and those preparing for examinations. I have included most of the mathematical and general constants as well as the tables normally required by those for whom the book is designed. Formulæ generally are scattered throughout text-books dealing with the various branches of mathematics, but being brought together in this convenient vest-pocket form the student will find the volume handier to have at hand for reference purposes wherever he is pursuing his studies.

The book will also be found of value by draughtsmen and technicians generally. It is uniform in size and style with my companion works—*Newnes Engineer's Pocket Book*, *Radio Engineer's Pocket Book*, *Wire and Wire Gauges*, and *Screw Thread Tables*.

F. J. CAMM.

PREFACE TO SIXTH EDITION

The call for a sixth edition of this book has enabled me to revise the text, and to correct minor printers' errors. I shall be glad to receive suggestions from readers regarding tables or formulæ that might be included in future editions.

F. J. CAMM.

CONTENTS

STANDARD MATHEMATICAL SYMBOLS

$+$ Plus, or add.
$-$ Minus, or subtract.
$\times$ Multiply by.
$\div$ Divide by.
$/$ Divide by.
$=$ Is equal to
$\equiv$ Is always equal to. Identical with.
$\simeq$ Approximately equal to.
$\doteqdot$ or $\doteq$ Approximately equal to.
$\therefore$ Therefore.
$\because$ Since, because.
(Single bracket.
{ Double bracket, or brace.
[Square bracket.
$\backsim$ Difference of.
$<$ Less than.
$>$ Greater than.
$\leqq$ Equal to, or less than.
$\geqq$ Equal to, or greater than.
$\nless$ Not less than.
$\ngtr$ Not greater than.
$\propto$ Varies as.
∞ Infinity.
$\parallel$ Parallel with.
$\perp$ Perpendicular to.
$-$ Vinculum or bar (but the use of brackets is preferable).
$\int$ Sign of differentiation.
$\pm$ Plus or minus, i.e., either plus or minus, according to circumstances.
$+\!\!+$ Modified plus sign, indicates that direction is taken into account as well as addition as in obtaining the vector sum of two forces.
$\forall$ Sign of vector subtraction.
Σ Sigma, the sum, or "summation of the products of."
π Pi, the ratio of circumference to diameter, also 180° in circular measure.
θ Theta, any angle from the horizontal.
Φ Phi, any angle from the vertical.
$\odot$ Circle, or station point. $\triangle$ Triangle, or trig station.
$\sqrt{}$ The radical square root. $\sqrt[3]{}$ Cube root.
$\sqrt[4]{}$ Fourth root.
$\underline{|5}$ means continued product up to $5=1\times2\times3\times4\times5$.
Πn = continued product of numbers up to n $=1\times2\times3\ .\ .\ .\ .\ .$ n.
a, *b*, *c* used for known quantities; *x*, *y*, *z* for unknown quantities. n is used in place of any whole number.

STANDARD MATHEMATICAL SYMBOLS (Continued).

A full stop (.) is sometimes used instead of the multiplication sign.

≠ Unequal to.

: As; so is (ratio).

:: Is to.

∦ Not parallel.

∠ Angle.

∟ Right angle.

▭ Parallelogram.

□ Square.

○ Circumference.

◠ Semicircle.

◔ Quadrant.

⌒ Arc.

(), [], { } Vincula.

THE GREEK ALPHABET

The Greek alphabet is as follows:

Α α (alpha), Β β (bēta), Γ γ (gamma), Δ δ (delta), Ε ε (epsilon), Ζ ζ (zēta), Η η (ēta), Θ θ (thēta), Ι ι (iōta), Κ κ (kappa), Λ λ (lambda), Μ μ (mu), Ν ν (nu), Ξ ξ (xi), Ο ο (ōmicron), Π π (pi), Ρ ρ (rho), Σ σ (sigma), Τ τ (tau), Υ υ (ūpsilon), Φ φ (phi), Χ χ (chi), Ψ ψ (psi), Ω ω (ŏmĕga).

ROMAN NUMERALS

1	2	3	4	5	6	7	8	9	10	50	100	500	1000
I	II	III	IV	V	VI	VII	VIII	IX	X	L	C	D	M

It will be observed that IV = 4, means 1 short of 5; in the same way IX = 9, means 1 short of 10; XL = 40, means 10 short of 50; XC = 90, means 10 short of 100; so for 1814 we have MDCCCXIV, and for 1957 we have MCMLVII

MATHEMATICAL AND GENERAL CONSTANTS

1 foot-pound = 1.3825×10^7 ergs.

1 horse-power = 33,000 ft.-lb. per min. = 746 watts.

A column of water 2.3 ft. high is equivalent to a pressure of 1 lb. per sq. in.

1 sidereal second = 0.99727 secs. (mean solar).

The length of seconds pendulum = 39.14 in.

Joule's Equivalent to suit Regnault's H

is $\begin{cases} 778{\cdot}1 \text{ ft.-lb.} = 1 \text{ Fahr. unit.} \\ 1{,}393 \text{ ft.-lb.} = 1 \text{ Cent. unit.} \end{cases}$

Regnault's H = 606.5 + .305 t° C. = 1,082 + .305 t° F.

pu 1·0646 = 479.

$$\log_{10} p = 6.1007 - \frac{B}{t} - \frac{C}{t^2}$$

where $\log_{10}$ B = 3.1812, $\log_{10}$ C. = 5.0882.

p is in pounds per sq. in., t is absolute temperature Centigrade.

u is the volume in cu. ft. per pound of steam.

Volts × amperes = watts.

1 electrical unit = 1,000 watts-hours.

1 second (mean solar) = 1·002738 sidereal secs.

MATHEMATICAL AND GENERAL CONSTANTS—continued

$\int_0^x$ is the sign of integration between limits 0 and x

The symbol $\int_a^b y.dx$ means "the area beneath the curve whose ordinate is y, from x = a to x = b."

$\Sigma \sin^3 \theta \int_{0°}^{90°}$ means the summation of the cubes of the sines of the angles 0 to 90.

The sign + (plus) or − (minus) has a higher power in a formula than × (multiplication), therefore parts connected by × may be multiplied out before passing beyond the other signs. Remember, that plus times plus = +, plus times minus = −, minus times minus = +.

e = Base of Napierian (named after Napier, discoverer of logarithms) or natural hyperbolic Logarithms = 2·7182818

To convert Napierian logarithms to common logarithms multiply by μ = (·4343).

To convert common to Napierian logs. multiply by $\frac{1}{\mu}$ = 2·30258509, or 2·3026 approx.

π = 3·14159265358979323846 . . . .
or 3·14159 approx. or 3·1416
or 3·142 or $3\frac{1}{7}$ = 3·142857.

π correct to six decimal places = $\frac{355}{113}$

Log π = 0·49714987269413385435

The base of Napierian or hyperbolic logarithms is the sum of the series:

$$2+\frac{1}{2}+\frac{1}{2\times3}+\frac{1}{2\times3\times4}+\frac{1}{2\times3\times4\times5}+\ldots$$

g = 32·182 f/s^2 = 980·9 cms./s^2. = 32·2 approx.
1 gallon = ·1606 cu. ft. = 10 lbs. of water at 62° F.
1 pint of water = $1\frac{1}{4}$ lbs.
1 cu. ft. of water weighs 62·3 lbs. (1,000 oz. approx.).
1 cu. ft. of air at 0° C. and 1 atmosphere weighs ·0807 lbs.

1 radian = $\frac{180}{\pi}$ = 57·2958 degrees.

π radians = 180°. $\frac{\pi}{2}$ radians = 90°. = 1·5707963 radians.

SOME FORMULÆ FREQUENTLY USED

APPROXIMATION

$(1\pm a)^2=1\pm 2a$

$(1\pm a)^3=1\pm 3a$

$\frac{1}{1\pm a}=1\mp a$

$\frac{1}{(1\pm a)^2}=1\mp 2a$

$\frac{1}{(1\pm a)^3}=1\mp 3a$

$\frac{1+a}{1-a}=1+2a$

$\frac{1-a}{1+a}=1-2a$

$\sqrt{1\pm a}=1\pm \frac{1}{2}a$

$\sqrt[3]{1\pm a}=1\pm \frac{1}{3}a$

a=small quantity.

TRIGONOMETRY

$\sin^2 A+\cos^2 A=1$

$\mathrm{cosec}^2 A=1+\cot^2 A$

$\cos A/2=\sqrt{\frac{s(s-a)}{bc}}$

$\sin A/2=\sqrt{\frac{(s-b)(s-c)}{bc}}$

$\tan A/2=\sqrt{\frac{(s-b)(s-c)}{s(s-a)}}$

$\frac{\sin A}{a}=\frac{\sin B}{b}=\frac{\sin C}{c}=\frac{1}{2R}$

$\tan\frac{B-C}{2}=\frac{b-c}{b+c}\cot A/2$

$c^2=a^2+b^2-2ab\cos C$

$1+\tan^2 A=\sec^2 A$

CALCULUS

$\frac{d}{dx}(ae^x)=ae^x$ $\qquad$ $\int ae^x dx=ae^x$

$\frac{d}{dx}(ax^n)=anx^{n-1}$ $\qquad$ $\int ax^n dx=\frac{a}{n+1}x^{n+1}$

$\frac{d}{dx}(a\log_n x)=\frac{a}{x}\log_n e$ $\qquad$ $\int \frac{a}{x}dx=a\log_e x$

$\frac{d}{dx}(a\sin bx)=ab\cos bx$ $\qquad$ $\int a\sin bx\,dx=-\frac{a}{b}\cos bx$

$\frac{d}{dx}(a\cos bx)=-ab\sin bx$ $\qquad$ $\int a\cos bx\,dx=\frac{a}{b}\sin bx$

$\frac{d}{dx}(a\tan bx)=ab\sec^2 bx$ $\qquad$ $\int a\tan bx\,dx=\frac{a}{b}\log\sec bx$

The formulæ in the first column remain true when x is replaced by $x+C$, where C is any constant.

A USEFUL SERIES

$$e^x = 1 + x + \frac{x^2}{2!} + \frac{x^3}{3!} + \frac{x^4}{4!} + \ldots .$$

$$\log_e \frac{m}{n} = 2\left[\left(\frac{m-n}{m+n}\right) + \frac{1}{3}\left(\frac{m-n}{m+n}\right)^3 + \frac{1}{5}\left(\frac{m-n}{m+n}\right)^5 + \ldots .\right]$$

$$\log_e x = 2\left[\left(\frac{x-1}{x+1}\right) + \frac{1}{3}\left(\frac{x-1}{x+1}\right)^2 + \frac{1}{5}\left(\frac{x-1}{x+1}\right)^5 + \ldots .\right]$$

$$\sin \theta = \theta - \frac{\theta^3}{3!} + \frac{\theta^5}{5!} - \frac{\theta^7}{7!} + \ldots .$$

$$\cos \theta = 1 - \frac{\theta^2}{2!} + \frac{\theta^4}{4!} - \frac{\theta^6}{6!} + \ldots .$$

$$\theta = \tan\theta - \frac{\tan^3\theta}{3} + \frac{\tan^5\theta}{5} - \frac{\tan^7\theta}{7} + \ldots .$$

ARITHMETICAL PROGRESSION

The term arithmetical progression refers to a series of numbers which increase or decrease by a constant difference. Thus: 1, 3, 5, 7, 9, or 10, 8, 6, 4 2, are arithmetical progressions, the constant difference being 2 in the first series and − 2 in the second.

Let a = the first term of the series, z = the last term, n = the number of terms, d = the constant difference, S = the sum of all the terms.

$a = z - d(n-1)$. $a = \frac{2S}{n} - z$. $a = \frac{S}{n} - \frac{d}{2}(n-1)$.

$z = a + d(n-1)$.

$z = \frac{2S}{n} - a$. $z = \frac{S}{n} + \frac{d}{2}(n-1)$. $n = \frac{z-a}{d} + 1$.

$n = \frac{2S}{a+z}$.

$d = \frac{z-a}{n-1}$. $d = \frac{(z+a)(z-a)}{2S-a-z}$. $d = \frac{2(S-an)}{n(n-1)}$

$d = \frac{2(zn-S)}{n(n-1)}$.

$S = \frac{n(a+z)}{2}$. $S = \frac{(a+z)(z+d-a)}{2d}$. $S = n[a + \frac{d}{2}(n-1)]$

$S = n\left[z - \frac{d}{2}(n-1)\right]$. $a = \frac{d}{2} \pm \sqrt{(z+\frac{d}{2})^2 - 2dS}$.

$z = -\frac{d}{2} \pm \sqrt{(a-\frac{d}{2})^2 + 2dS}$. $n = \frac{1}{2} - \frac{a}{d} \pm \sqrt{(\frac{1}{2} - \frac{a}{d})^2 + \frac{2S}{d}}$

$n = \frac{1}{2} + \frac{z}{d} \pm \sqrt{\left(\frac{1}{2} + \frac{z}{d}\right)^2 - \frac{2S}{d}}$

When the series is decreasing make the first term = z, and the last term = a. The Arithmetical Mean of two quantities, A and B $= \frac{A+B}{2}$

GEOMETRICAL PROGRESSION

A geometrical progression refers to a series of numbers which increase or decrease by a constant factor, or common ratio. For example, 2, 4, 8, 16, 32; 4, $-\frac{1}{2}$, 1-16, $-$ 1-128, 1-1024, are Geometrical Progressions, the constant factor being 2 in the first series and $-\frac{1}{8}$ in the second. Let a = the first term, z = the last term, n = the number of terms, r = the constant factor and S = the sum of the terms.

$a=\frac{z}{r^{n-1}}$. $a=S-r(S-z)$. $a=S\frac{r-1}{r^n-1}$. $z=ar^{n-1}$.

$z=S-\frac{S-a}{r}$. $z=S\left(\frac{r-1}{r^n-1}\right)r^{n-1}$. $r=\sqrt[n-1]{\frac{z}{a}}$.

$r=\frac{S-a}{S-z}$.

$ar^n+S-rS-a=0$. $S=a\frac{(r^n-1)}{r-1}$. $S=a\frac{(1-r^n)}{1-r}$.

$S=\frac{rz-a}{r-1}$.

$S=\frac{z\,(r^n-1)}{(r-1)r^{n-1}}$ $n=1+\frac{\log z-\log a}{\log r}$.

$n=1+\frac{\log z-\log a}{\log(S+a)-\log(S-z)}$.

$n=\frac{\log[a+S(r-1)]-\log a}{\log r}$.

$n=1+\frac{\log z-\log[zr-S(r-1)]}{\log r}$.

$S=\frac{z^{n-1}\sqrt{z}-a^{n-1}\sqrt{a}}{{}^{n-1}\sqrt{z}-{}^{n-1}\sqrt{a}}$.

The Geometric Mean of two quantities, A and B $=\sqrt{AB}$

HARMONICAL PROGRESSION

Quantities are said to be in Harmonical Progression when, any three consecutive terms being taken, the first is to the third as the difference between the first and second is to the difference between the second and third. Thus, if a, b, c, be the consecutive terms in a series, then, if $a : c :: a - b : b - c$, a, b, c are in harmonical progression. If quantities are in harmonical progression, their

HARMONICAL PROGRESSION—Contd.

reciprocals must also be in arithmetical progression.

The Harmonic Mean of two quantities, A and B $= \frac{2AB}{A+B}$

EXTRACTING SQUARE ROOT

Mark off the number, the square root of which is to be found, into periods by marking a dot over every second figure commencing with the units place. Draw a vertical line to the left of the figure and a bracket on the right-hand side. Next, find the largest square in the left-hand period, and place this root behind the bracket. Next, the square of this root is subtracted from the first period, and the next period is brought down adjacent to the remainder and used as a dividend. Now, multiply the first root found by 2 and place this product to the left of the vertical line; then divide it into the left-hand figures of this new dividend, ignoring the right-hand figure. Attach the figure thus obtained to the root, and also to the divisor. Multiply this latest divisor by the figure of the root last obtained, finally subtracting the product from the dividend. Continue this operation until all periods have been brought down. If a decimal fraction is involved the periods for the decimal are marked off to the right of the decimal point. The following examples will make the process clear. The first trial divisors are underlined in each case.

Example.

Find the square root of 1156:

```
 3 | 1156(34
   |  9
64 |  256
   |  256
```

Find the square root of 54756:

```
  2 | 54756(234
    | 4
 43 | 147
    | 129
464 |  1856
    |  1856
```

In dealing with decimals, the periods relating

EXTRACTING SQUARE ROOT—
(Continued)

to the decimal are marked off to the right as previously mentioned.

Find the square root of 39·476089 :

Divisor	Dividend
6	39·476089(6·283
	36
122	347
	244
1248	10360
	9984
12563	37689
	37689

.....

EXTRACTING CUBE ROOT

Separate the given number into periods by placing a dot over every third figure above the decimal point.

Determine the greatest number whose cube does not exceed the number in the left-hand period. This number is the first part of the required cube root. It is here called a " partial root." Subtract its cube from the left-hand period and to the remainder bring down the next period to form a second dividend. To form the next divisor, first form a partial divisor by squaring the partial root, multiplying by 3 and adding two ciphers. The addition to be made to this partial root is something less than the quotient of the second dividend and the partial divisor. It has to be found by trial and is here called the " trial addition to the root." The partial divisor is increased by—

(*a*) 3 times the product of the partial root (with one cipher added) and the trial addition to the root; (*b*) the square of the trial addition to the root.

This gives the second divisor, which is then multiplied by the trial addition to the root. If the product is greater than the dividend the trial addition is too large.

If the dividend exceeds the product by an amount greater than the divisor the trial addition is too small.

When a trial addition has been found to avoid these extremes, the next period from the original number is added to the remainder and the work proceeds as before.

PERMUTATIONS AND COMBINATIONS

The number of arrangements or permutations of n different things taken r at a time is written ^nP_r, which is the product of r factors beginning with n and decreasing by 1 each time.

$^nP_r = n(n-1)(n-2)(n-3)\ \ldots$ to r factors.

The last factor will be $n-(r-1)$, or $n-r+1$.

Therefore :

$^nP_r = n(n-1)\,(n-2)\ \ldots\ (n-r+1)$.

The number of permutations of n different things arranged in a row is :

$n(n-1)\,(n-2)\ \ldots\ 3,\ 2,\ 1 = \underline{|n}$.

It is convenient for factorial $\underline{|0}$ to be considered as 1.

The number of combinations of n things taken r at a time is ^nC_r.

Therefore :

$$^nC_r = \frac{^nP_r}{\underline{|r}} = \frac{n(n-1)(n-2)\ \ldots\ (n-r+1)}{\underline{|r}}$$

It is often more convenient to use

$$^nP_r = \frac{\underline{|n}}{\underline{|n-r}}$$

Hence:

$$^nC_r = \frac{\underline{|n}}{\underline{|r}\ \underline{|n-r}}$$

The difference between permutations and combinations is: in permutations the order of things is taken into account, but in combinations the order is of no account.

Hence, xyz, xzy, yxz, yzx, are different permutations of the same combinations.

SIMPLE INTEREST

A =amount of principal+interest.
I =interest.
R =rate per cent.
P =principal.
T =time in years.

$$I = \frac{PTR}{100}$$

$$P = \frac{100\ I}{RT}$$

$$R = \frac{100\ I}{PT}$$

$$T = \frac{100\ I}{PR}$$

$$P = \frac{100\ A}{100+RT}$$

$$A = P+I = \frac{P\ (100+RT)}{100}$$

If the time is given in days, weeks or months, multiply the 100 in these equations by 365, 52, or 12 as the case may be.

COMPOUND INTEREST

P =principal.
A =amount.
n =number of years.
r =interest on £1 for a year.

$$P = \frac{A}{(1+r)^n}$$

$$n = \frac{\text{Log } A - \text{Log } P}{\text{Log}(1+r)}$$

$$A = P(1+r)^n$$

$$r = n\sqrt{\frac{A}{P}} \quad 1 = \sqrt[n]{\frac{A}{P}} - 1$$

PRESENT VALUE AND DISCOUNT

The present value or present worth of a sum of money due at some future date is equal to the sum of money which, plus its interest from the present time to the time when it becomes due for payment will be equal to the given amount. Thus, the present value of £412 in 9 months time at 4 per cent. simple interest is £400.

True discount is the difference between the amount and the present value—the interest on the present value for a given period.

Banker's discount is interest on the amount of the debt, and thus is always greater than the true discounts.

In the example given the banker's discount would be £12 7s. $2\frac{2}{5}$d., whereas the true discount is £12.

A = amount of debt.
D = true discount.
D_1 = banker's discount.
P = present value.
r = interest on £1 for one year.
n = number of years.

$$A = \frac{D(1+nr)}{nr}$$

$$D = A-P = A - \frac{A}{1+nr} = \frac{A\,nr}{1+nr}$$

$$A = P(1+nr).$$

$$D_1 = Anr$$

$$r = \frac{A-P}{Pn}$$

$$n = \frac{A-P}{Pr}$$

$$P = \frac{A}{1+nr}$$

ANNUITIES

A fixed sum of money paid periodically is an annuity. The present value of an annuity which is to continue for n years at compound interest is:

$$P = \frac{1-(1+r)^{-n}}{r}N$$

In the case of a perpetual annuity the present value is found by making n infinite in this formula.

Hence in this case:

$$P = \frac{n}{R}$$

If an annuity is left unpaid for n years the whole amount A at simple interest (r=interest of £1 for one year) is:

$$A = nN + \frac{n(n-1)}{2}rn$$

ANNUITIES—(*Continued*)

At compound interest

$$A=\frac{(1+r)^{n}-1}{r}N$$

It is not possible to calculate with any degree of reliability the present value of an annuity at simple interest.

Let mN = present value of an annuity.
N = annuity.

Then the annuity is worth m years' purchase.

The number of years' purchase (m) of a perpetual annuity is:

$$m=\frac{1}{r}=\frac{100}{\text{rate per cent.}}$$

In other words, an annuity is worth m year's purchase.

A debt which bears compound interest is reduced by an annual payment N.

The formula for finding the debt d after n years, and the time m when it will be fully paid is:

$$d=\frac{(Dr-N)\ (1+r)^{n}+N}{r}$$

$$m=\frac{\text{Log } N-\text{Log } (N-Dr)}{\text{Log } (1+r)}$$

When $n = Dr$ the debt will never be paid.

To find the annual payment N which will clear a debt D at r per cent. compound interest in n years:

$$N=\frac{Dr\ (1+r)^{n}}{(1+r)^{n}-1}$$

EULER'S NUMBERS E_n

$2/(e^{x}+e^{-x})=1+\Sigma(-1)^{n}E_{n}x^{2n}/(2n)!$

$E_1=1$

$E_2=5$

$E_3=6.1\times10$

$E_4=1.385\times10^{3}$

$E_5=5.0521\times10^{4}$

$E_6=2.702765\times10^{6}$

$E_7=1.99360981\times10^{8}$

$E_8=1.9391512145\times10^{10}$

$E_9=2.404879675441\times10^{12}$

EULER'S CONSTANT

$$\gamma=\underset{n=\infty}{\text{Lt}}\left(\frac{1}{1}+\frac{1}{2}+\frac{1}{3}+\ \ldots\ +\frac{1}{n}-\log_{e} n\right)$$

$=.5721$

BINOMIAL THEOREM

Binomial theorem is $(1+x)^n$ = a certain expansion.

$$(1+x)^n=1+{}^nC_1x+{}^nC_2x^2+{}^nC_3x^3+\ \ldots\ +{}^nC_nx^n$$

or

$$(1+x)^n=1+\frac{n}{\lfloor 1}x+\frac{n(n-1)}{\lfloor 2}x^2+\frac{n(n-1)(n-2)}{\lfloor 3}x^3+$$

$$.+x^n$$

$$x^r={}^nC_rx^r \text{ or } \frac{n(n-1)(n-2)\ldots(n-r+1)}{\lfloor r}x^r$$

This applies only for a positive integral index.

ALGEBRAIC IDENTITIES

$$(a+b)^2=a^2+2ab+b^2$$

$$(a-b)^2=a^2-2ab+b^2$$

$$(a+b)(a-b)=a^2-b^2$$

$$a^3+b^3\equiv(a+b)(a^2-ab+b^2)$$

$$a^3-b^3\equiv(a-b)(a^2+ab+b^2)$$

$$(a+b)^3\equiv a^3+3a^2b+3ab^2+b^3$$

$$(a-b)^3\equiv a^3-3a^2b+3ab^2-b^3$$

$$a^3+b^3+c^3-3abc\equiv(a+b+c)(a^2+b^2+c^2-bc-ca-ab)$$

AREA OF IRREGULAR FIGURES

A reliable method of calculating the area of irregular figures, such as the diagram A B D C, is given below.

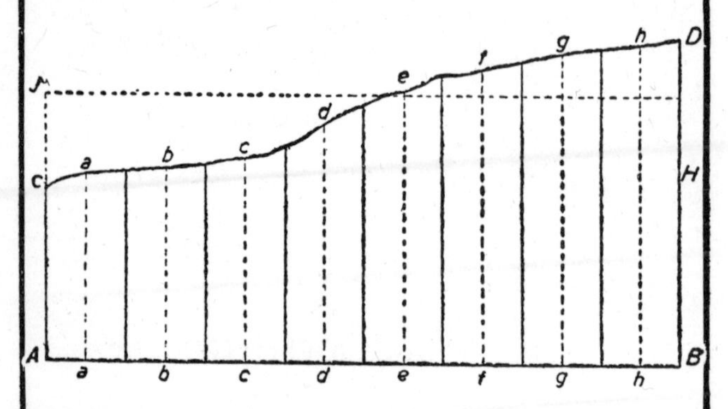

Divide the baseline AB into any convenient number of equal parts. The greater the number the more accurate the result. At the centre of each part erect ordinates, as shown dotted. Now measure the height of the ordinates aa, bb, cc, dd, ee, ff, gg, hh, and add them together. Divide their sum by the number of ordinates. This will give the mean height H, or mean ordinate AJ. A simple method of obtaining the total length of the ordinates is to use the edge of a piece of paper, starting with aa and adding bb, cc, etc. The area is then found by multiplying AB by AJ.

The Trapezoidal Method.—Divide the base into a number of equal parts (say 8), spaced a distance *s* apart. Let the height of each ordinate be h_1, h_2, h_3, h_4, etc., and the mean height h. Then:

$$h=\frac{s}{8}\left\{\tfrac{1}{2}(h_1+h_9)+h_2+h_3+h_4+h_5+h_6+h_7+h_8\right\}$$

(The number of ordinates is always 1 more than the number of parts.) Expressed as a rule:

AREA OF IRREGULAR FIGURES—
(*Continued*)

Divide the base into any number of equal parts, and add ½ the sum of the end ordinates to the sum of all the others. Multiply the result by the common interval s *to obtain the area; or divide by the number of spaces to obtain the mean ordinate.*

Multiply the mean ordinal by the length to obtain the area.

Simpson's Rule.—Probably the most accurate method. Divide the base AB into an even number of equal parts (say, 6) to produce an odd number of ordinates (7), spaced a distance s apart. Then:

Area ABDJ

$$=\frac{s}{3}\left\{h_1+h_7+4(h_2+h_4+h_6)+2(h_3+h_5)\right\}$$

This reduces to:

$$\frac{s}{3}(A+4B+2C),$$

where A = sum of 1st and last ordinates.
B = sum of even ordinates.
C = sum of odd ordinates.

Expressed as a rule:

Add together the extreme ordinates, 4 times the sum of the even ordinates, and twice the sum of the odd ordinates (omitting the first and last); then multiply the result by one third the space between the ordinates.

In the case of an irregular figure in which the end ordinates are zero, then A is zero, and the formula then becomes:

$$\frac{s}{3}(0+4B+2C)$$

In some cases, where the irregular figure is bounded by two curved lines, it is convenient to divide the figure into two parts, and calculate the area of each independently.

When the area is completely bounded by an irregular curve, parallel lines are drawn, touching the top and bottom of the curve, and vertical lines touching the sides. The figure is thus enclosed in a rectangle, and by means of ordinates the area can be found as before.

MENSURATION

A & a=area; b=base; C & c=circumference; D & d=diameter; h=height; n°=number of degrees; p=perpendicular; R & r=radius; s=span or chord; v=versed sine.

SQUARE

$a = \text{side}^2$; $\text{side} = \sqrt{a}$; $\text{diagonal} = \text{side} \times \sqrt{2}$.

Side × 1.4142 = diameter of circumscribing circle.

Side × 4.443 = circumference of circumscribing circle.

Side × 1.128 = diameter of circle of equal area.

Area in square inches × 1.273 = area of equal circle.

CIRCLE

$a = \pi r^2 = d^2\frac{\pi}{4} = .7854d^2 = .5cr$; $c = 2\pi r$, or $\pi d = 3.14159d$, or $3.1416d$ approx. $= 3.54\sqrt{a} = 3\frac{1}{7}d$ or $\frac{22}{7}d$ approx.

Side of equal square = .8862d.

Side of inscribed square = .7071d.

d = .3183c.

Circle has maximum area for given perimeter.

ANNULUS

$a = (D+d)(D-d)\frac{\pi}{4} = \frac{\pi}{4}(D^2-d^2)$

SEGMENT OF CIRCLE

a = area of sector − area of triangle

$= \frac{4v}{3}\sqrt{(0.625v)^2 + (\frac{1}{2}s)^2}$

LENGTH OF ARC

Length of arc $=.0174533n°r$

$$=\frac{1}{3}\left(8\sqrt{\frac{s2}{4}+v2-s}\right)$$

Approx. length of arc $=\frac{1}{3}$(8 × chord of ½ arc − chord of whole arc).

$$d=\frac{(\frac{1}{2}\text{ chord})}{v}+v$$

radius of curve $=\frac{s2}{8v}+\frac{v}{2}$

SECTOR OF CIRCLE

$a=.5r \times$ length of arc $=\frac{n°}{360}\times$ area of circle

RECTANGLE OR PARALLELOGRAM

$$a=bp$$

TRAPEZIUM

(Two Sides Parallel)

a = mean length of parallel sides × distance between them.

ELLIPSE

$a=\frac{\pi}{4}Dd=\pi Rr$; $c=\sqrt{\frac{D2+d2}{2}}\times\pi$ approx. or $\pi\frac{Da}{2}$ approx.

PARABOLA

$$a = \frac{2}{3}bh$$

CONE OR PYRAMID

$$\text{Surface} = \frac{\text{circum. of base} \times \text{slant length}}{2}$$
$$+ \text{area of base.}$$

Volume = area of base × $\frac{1}{3}$ vertical height.

FRUSTUM OF CONE

Surface = (C + c) × $\frac{1}{2}$ slant height + area of ends.

$$\text{Volume} = .2618h\,(D^2 + d^2 + Dd)$$
$$= \tfrac{1}{3}h\,(A + a + \sqrt{A \times a})$$

PRISM

Volume = area of base × height.

WEDGE

$$\text{Volume} = \frac{1}{6}(\text{length of edge} + 2 \text{ length of back})bh.$$

PRISMOIDAL FORMULA

The prismoidal formula enables the volume of a prism, pyramid, or frustum of a pyramid to be found.

$$\text{Volume} = \frac{\text{end areas} + 4 \times \text{mid area}}{6} \times \text{height.}$$

SPHERE

$$\text{Surface} = d^2\pi = 4\pi r^2$$

$$\text{Volume} = \frac{d^3\pi}{6} = \frac{4}{3}\pi r^3$$

SEGMENT OF SPHERE

r = rad. of base.

Volume $= \frac{\pi}{6} h(3r^2 + h^2)$.

r = rad. of sphere.

Volume $= \frac{\pi}{3} h^2(3r - h)$.

CUBE

Volume = length × breadth × height.

SPHERICAL ZONE

Volume $= \frac{\pi}{2} h(\frac{1}{3}h^2 + R^2 + r^2)$.

Surface area of convex part of segment or zone of sphere $= \pi d$(of sphere)h

$= 2\pi rh$.

Mid-spherical zone:

Volume $= (r + \frac{2}{3}h^2)\frac{\pi}{4}$

SPHEROID

Volume = revolving axis^2 × fixed axis × $\frac{\pi}{6}$

CYLINDER

Area $= 2\pi r^2 + \pi dh$.

Volume $= \pi r^2 h$.

TORUS

Volume $= 2\pi^2 Rr^2$

$= 19.74 Rr^2$

$= 2.463 Dd^2$

FORMULÆ FOR TRIANGLES

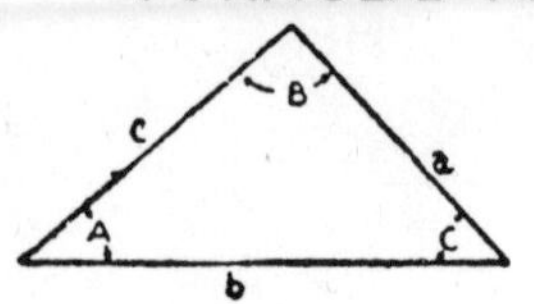

Fig. 1. Diagram for Table A.

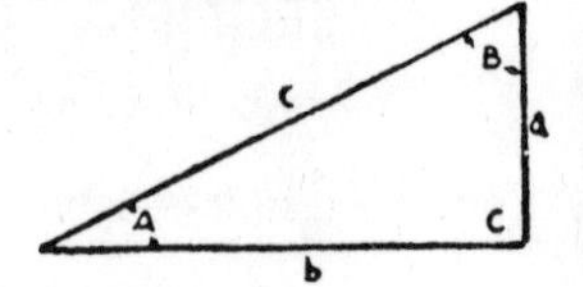

Fig. 2. Diagram for Table B.

TABLE A

See Fig. 1

Parts Given.	Parts to be Found.	Formulæ.
$a\ b\ c$	A	$\cos A = \frac{b^2+c^2-a^2}{2bc}$
$a\ b$ A	B	$\sin B = \frac{b \times \sin A}{a}$
$a\ b$ A	C	$C = 180° - (A+B)$
a A B	b	$b = \frac{a \times \sin B}{\sin A}$
a A B	c	$c = \frac{a \sin C}{\sin A} = \frac{a \sin (A+B)}{\sin A}$
$a\ b$ C	B	$B = 180° - (A+C)$

TABLE B

See Fig. 2.

Parts Given.	Parts to be Found.				
	A	B	a	b	c
a & c	$\sin A=\frac{a}{c}$	$\cos B=\frac{a}{c}$		$b=\sqrt{c^2-a^2}$	
a & b	$\tan A=\frac{a}{b}$	$\cot B=\frac{a}{b}$			$c=\sqrt{a^2+b^2}$
c & b	$\cos A=\frac{b}{c}$	$\sin B=\frac{b}{c}$	$a=\sqrt{c^2-b^2}$		
A & a		$B=90°-A$		$b=a\times\cot A$	$c=\frac{a}{\sin A}$
A & b		$B=90°-A$	$a=b\times\tan A$		$c=\frac{b}{\cos A}$
A & c		$B=90°-A$	$a=c\times\sin A$	$b=c\times\cos A$	

Fig. 3.—In any right-angled triangle:

$$\tan A=\frac{BC}{AC},\quad \sin A=\frac{BC}{AB}$$

$$\cos A=\frac{AC}{AB},\quad \cot A=\frac{AC}{BC}$$

$$\sec A=\frac{AB}{AC},\quad \mathrm{cosec}\ A=\frac{AB}{BC}$$

Fig. 3.

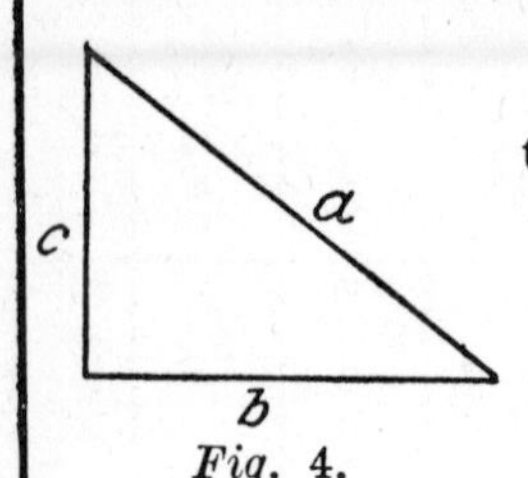

Fig. 4.

Fig. 4.—In any right-angled triangle :

$$a^2 = c^2 + b^2$$

$$c = \sqrt{a^2 - b^2}$$

$$b = \sqrt{a^2 - c^2}$$

$$a = \sqrt{b^2 + c^2}$$

Fig. 5.—$c + d : a + b :: b - a : d - c$.

$$d = \frac{c + d}{2} + \frac{d - c}{2}$$

$$x = \sqrt{b^2 - d^2}$$

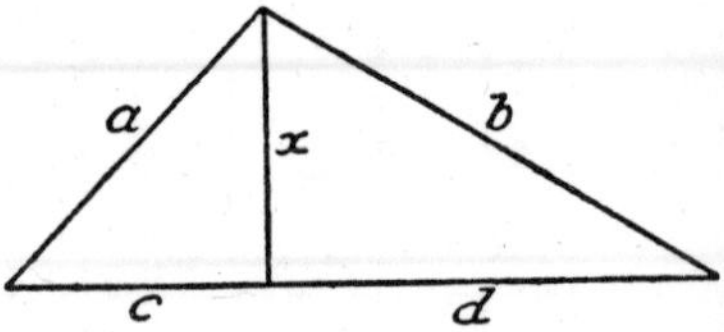

Fig. 5.

Fig. 6.

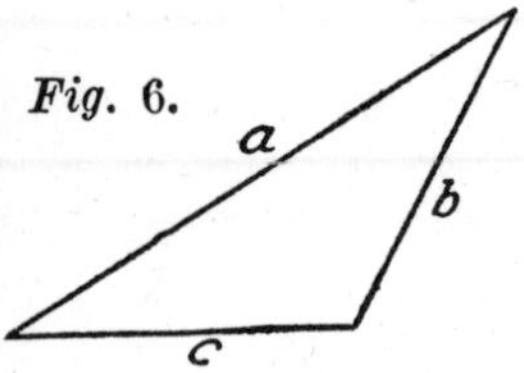

In Fig. 6, where the lengths of three sides only are known :

area $= \sqrt{s(s - a)\ (s - b)\ (s - c)}$

where $s = \dfrac{a + b + c}{2}$

Fig. 7.—In this diagram :

$a : b :: b : c$, or $\dfrac{b^2}{a} = C$.

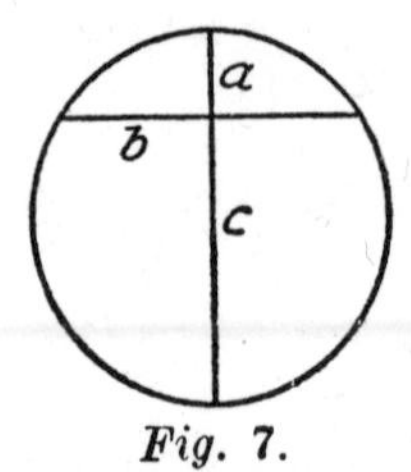

Fig. 7.

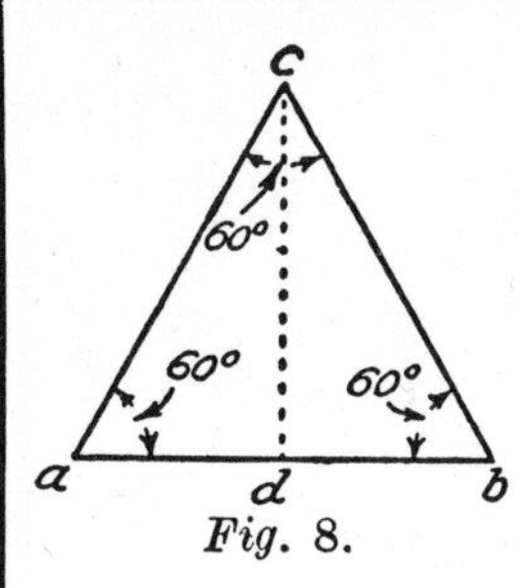

Fig. 8.

Fig. 8.—In an equilateral triangle $ab = 1$, then $cd = \sqrt{0.75} = 0.866$, and $ad = 0.5$; $ab = 2$, then $cd = \sqrt{0.3} = 1.732$, and $ad = 1$; $cd = 1$, then $ac = 1.155$ and $ad = 0.577$; $cd = 0.5$, then $ac = 0.577$ and $ad = 0.288$.

Fig. 9.—In a right-angled triangle with two equal acute angles, $bc = ac$. $bc = 1$, then $ab = \sqrt{2} = 1.414$; $ab = 1$, then $bc = \sqrt{0.5} = 0.707$.

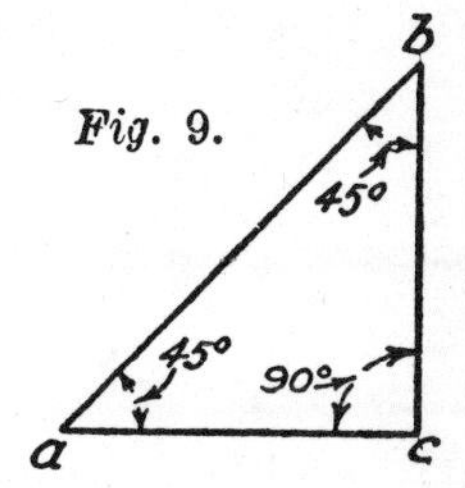

Fig. 9.

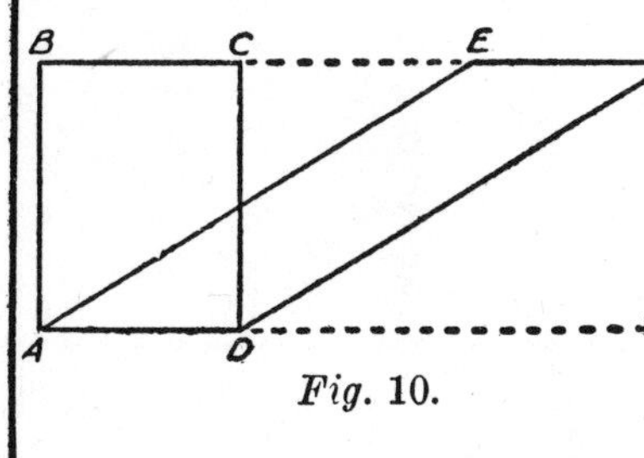

Fig. 10.

Fig. 10 shows that parallelograms on the same base and between the same parallels are equal; thus ABCD=ADEF.

Fig. 11 demonstrates that triangles on the same base and between the same parallels are equal in area; thus ABC = ADC.

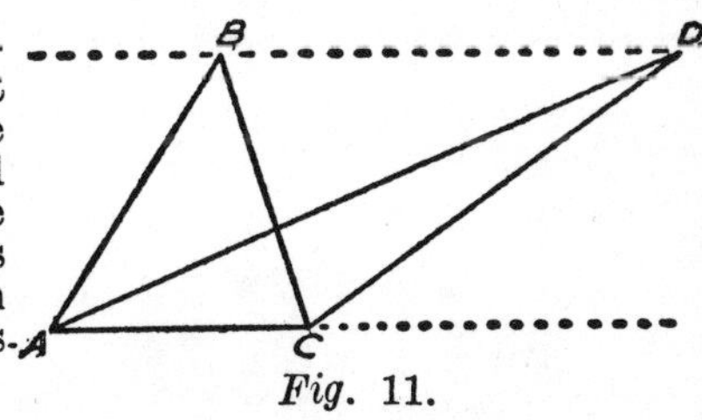

Fig. 11.

TRIGONOMETRICAL EQUIVALENTS

Sine	=	$\sqrt{1 - \text{Cos}^2}$.	Cosecant	=	1 ÷ Sin.
Sine	=	1 ÷ Cosec.	Tangent	=	1 ÷ Cotan.
Sine	=	Cos ÷ Cotan.	Tangent	=	Sin ÷ Cos.
Sine	=	Tan ÷ Sec.	Cotangent	=	1 ÷ Tan.
Cosine	=	$\sqrt{1 - \text{Sin}^2}$.	Cotangent	=	Cos ÷ Sin.
Cosine	=	1 ÷ Sec.	Versine	=	1 — Cos.
Cosine	=	Sin × Cotan.	Coversine	=	1 — Sin.
Cosine	=	Sin ÷ Tan.	1	=	Tan × Cotan.
Secant	=	1 ÷ Cos.	1	=	$\text{Sin}^2 + \text{Cos}^2$.
Secant	=	Tan ÷ Sin.	Secant^2	=	$1 + \text{Tan}^2$.

TRIGONOMETRICAL FUNCTIONS

RIGHT-ANGLED TRIANGLES

(See Fig. 2.)

$\text{Tan A} = \frac{a}{b}$ $\text{Cos A} = \frac{b}{c}$ $\text{Sin A} = \frac{a}{c}$

$\text{Sec A} = \frac{c}{b}$ $\text{Cot A} = \frac{b}{a}$ $\text{Cosec A} = \frac{c}{a}$

$\text{Versin A} = \frac{c - b}{c}$ $\text{Coversin A} = \frac{c - a}{c}$

UNITS.

Velocity, Acceleration, Force, Energy and Power.

Derived Units.—Derived units are units for the measurement of quantities which are based upon the fundamental units.

Absolute Units.—These are derived units referred directly to the fundamental units of mass, length and time.

Fundamental Units.—The fundamental units are the units of length, mass and time.

Arbitrary Units.—These are arbitrary combined units, not related to the fundamental units; thus the weight of 1 pound is an arbitrary unit because it depends upon the locality of the earth and upon the size and density of the earth.

At any instant the linear velocity of a particle is the rate at which its position is changing. The unit of velocity is a rate of change of position of 1 ft. per second, or 1 centimetre per second. Velocity is expressed by length divided by time, and the dimensions of a velocity are therefore length divided by time. This is expressed by the dimensions equation $V = LT^{-1}$.

Acceleration is at the rate at which velocity is changing at any instant. The dimensions of acceleration are LT^{-2} or $\frac{L}{T^2}$ and are expressed in feet or centimetres per second per second. On the C.G.S. system, that is in centimetres and seconds, acceleration of gravity is 981 centimetres per second per second, and the acceleration of gravity in feet per second per second is $\frac{981}{30.5} = 32.2$.

Momentum is the product of mass and velocity at any instant. Its dimensions are therefore $\frac{ML}{T} = MLT^{-1}$.

Force is measured by the rate at which the momentum of a body is changing at the particular instant. The dimensions of force are $\frac{ML}{T^2}$ or MLT^{-2}.

Newton's Laws of Motion.—(1) A body cannot

UNITS.

Velocity, Acceleration, Force, Energy and Power.—(*Continued*).

change its own linear or angular momentum except by the application of some external force or torque. (2) Force is measured by the rate of change in momentum in the direction of the impressed force. (3) If any force acts on two bodies, the change in momentum in both of them is the same.

The Dyne.—The unit of force on the centimetre-gramme-second system is the dyne. It is the force which after acting on a mass of 1 gramme for 1 second gives it a velocity of 1 centimetre per second.

The Poundal.—The unit of force on the foot-pound system is the poundal. It is defined as the dyne, but with the pound and foot substituted for gramme and centimetre. The poundal is an absolute unit of force on the f.p.s. system, but the weight of 1 pound is a gravitation and arbitrary unit and varies with the locality.

Distinction must be made between linear velocity, acceleration, momentum and force, and angular velocity, acceleration, momentum and torque or angular force. In the linear quantities we have m or mass, and v which stands for linear velocity of the centre of mass or centre of gravity, but in the angular equivalent we have to substitute for mass the moment of inertia round the axis of rotation K, and for an element of path ds, an element of angle $d\theta$.

Moment of Inertia.—The moment of inertia round any axis is defined as the sum of each element of mass of the body, each multiplied by the square of its perpendicular distance from the axis, thus the angular velocity is $\frac{d\theta}{dt}$ The angular acceleration is $\frac{d^2\theta}{dt^2}$. The angular momentum is $\frac{d\theta}{Kdt}$ and the torque or angular force or impressed couple is $\frac{d^2\theta}{Kdt^2}$.

UNITS.

Velocity, Acceleration, Force, Energy and Power.— (*Continued*).

Energy or Work.—When a body is displaced against the action of a force tending to move it in the opposite direction, work is said to be done or energy expended on it. In the case of a mass m raised to height h against the uniform acceleration of gravity g work equal to mgh units is done.

The Foot-Pound.—The unit of work in mechanical engineering is the foot-pound, and it is the work done in lifting 1 pound 1 foot against gravity. This, however, is dependent on the acceleration of gravity at the locality.

The Foot-Poundal.—The absolute unit of work in the foot-pound system is the foot-poundal, which is nearly equal to $\frac{1}{32.2}$ foot-pounds.

The Centimetre-Dyne or Erg.—The unit of work on the C.G.S. system is a centimetre-dyne, which is called the erg. Ten million ergs are called 1 joule. The dimensions of work are those of force multiplied by length or ML^2T^{-2}.

The Horse-Power Hour.—This is a common engineering unit of work. 1 horse-power equals a rate of doing work of 33,000 foot-pounds per minute, or 550 foot-pounds per second. A horse-power hour is the amount of work done when 1 horse-power continues for 1 hour. It is therefore equal to 1,980,000 foot-pounds. It is therefore a gravitation unit and depends on locality. As 1 foot-pound is approximately equal to 32.2 foot-poundals, 1 horse-power hour is equal to 63.75 times 10^6 foot-poundals, which equals 1,980,000 times 32.2.

The dimensions of power are those of work divided by time, which equals ML^2T^{-3}. The change ration from f.p.s. to c.g.s. is therefore 421,390.

Power.—Power is the rate of doing work. The c.g.s. unit of power is therefore 1 erg per second. The f.p.s. unit of power is 1 foot-poundal per second.

HORSE POWER

The unit of work (Horse power) is based on the assumption that a horse can travel 2½ miles per hour for 8 hours a day, performing the equivalent of pulling a load of 150 lb. out of a shaft by means of a rope. Thus 2½ miles an hour is 220 ft. per minute, and at that speed the load of 150 lb. is raised vertically the same distance. Therefore 300 lb. would be raised 110 ft., or 3,000 lb. raised 11 ft., or 33,000 lb. raised 1 ft. high per minute. The latter is the unit of horse-power, i.e. 33,000 lb. raised 1 ft. high per minute, or 33,000 foot-lb. per minute. Electrical equivalent is 746 watts.

Horse Power of an Electric Motor

$$\text{I.H.P.} = \frac{\text{Volts} \times \text{Amperes}}{746}$$

Horse Power (Indicated) of a Steam Engine (Single-Acting)

$$\text{I.H.P.} = \frac{PLAN}{33{,}000}$$

Where P = Mean effective steam pressure in lb. per sq. in.
L = Length of stroke in feet.
A = Area of piston in sq. in.
N = Number of revolutions per minute.

For a double-acting engine the formula is:

$$\text{I.H.P.} = \frac{2PLAN}{33{,}000}$$

Horse Power of Petrol Engines

R.A.C. Formula: $\text{H.P.} = \frac{D^2N}{1613}$

Dendy Marshal Formula: $\text{H.P.} = \frac{D^2SNR}{200{,}000}$

Where S = Stroke in centimetres.
D = Diameter of cylinder in centimetres.
R = Revolutions per minute.
N = Number of cylinders.

A.C.U. Formula: 100 c.c. = 1 h.p.

ELECTRICAL UNITS

The volt is the practical unit of electromotive force, or difference of potential or electrical pressure.

The ohm is the practical unit of resistance, which varies directly as the length and inversely as the area of section of a conductor.

The ampere is the practical unit of strength of current, or velocity. It is often described as the measure of current density.

The coulomb is the practical unit of quantity, and represents the amount of electricity given by 1 ampere in 1 second. (The term "coulomb" is becoming obsolete.)

The farad is the unit of capacity of an electrical condenser; one-millionth of this, or the micro-farad, is taken as the practical unit. A condenser of 1 farad capacity would be raised to the potential of 1 volt by the charge of 1 coulomb of electricity.

The watt is the practical unit of work, and is the amount of work required to force 1 ampere through 1 ohm during 1 second.

The joule is the mechanical equivalent of heat, or the work done in generating one heat unit or caloric, and is equal to 1 watt per second. The watt-hour joule $=10^7$, or 100,000,000 ergs. The Board of Trade Unit equals 3,600,000 joules, and is known as the Kilowatt-hour. Named after Joule, the English physicist. Joule's Law states that the heat produced by a current I passing through a resistance R for time t is proportional to I^2Rt; 1 calorie $=4.2 \times 10^7$ ergs, or 4.2 joules.

The henry is the unit of inductance or the coefficient of self-induction.

The volt may be understood as a measure of pressure, the ampere of quantity, the watt of power; thus a current of 10 amperes at 10 volts = 100 watts.

The Board of Trade Commercial Unit = 1 kelvin = 3,600,000 joules = 1,000 watt-hours = 1 kilowatt per hour = 1.34 H.P. for 1 hour. It equals a current of 1,000 amperes at an E.M.F. of 1 volt flowing for 1 hour. For private lighting it may be taken as 10 A 100 V and for public lighting 5 A 200 V.

ELECTRICAL UNITS—Continued.

Measure of Electrical Work

A = strength of current in amperes.
V = electromotive force in volts.
O = resistance in ohms.
C = quantity of electricity in coulombs.
t = time in seconds.
H.P. = actual horse-power.
W = units of work or watts (1 unit = 10 million ergs = absolute C.G.S. measurement).

$$A = \frac{V}{O} = \frac{C}{t}, \quad C = At.$$

$$H.P. \frac{A\ V}{746} = \frac{A^2 O}{746} = \frac{W}{746}.$$

$$W = A\ V = A^2 O.$$

1 watt $\frac{1}{746}$ of a H.P. = 1 volt ampere = 10^7 ergs per second.
1 kilowatt = 1,000 watts = 10^{10} ergs per second.
1 kilowatt-hour = 1.34 H.P. acting for 1 hour = say $2\frac{3}{4}$ million foot-lb.

Electrical Equations

Amperes × volts = watts.
Joules ÷ seconds = watts.
Coulombs per second = amperes.
Watts ÷ 746 = effective H.P.
Coulombs ÷ volts = farads.
0.7373 foot-lb. per second = 1 joule.
Volts × coulombs = joules.
Watts × 44.236 = foot-lb. per minute.
Kilowatts × 1.34 = H.P.

An erg is the work done by 1 dyne acting through 1 centimetre. A dyne is $\frac{1}{981}$ of a gramme. A gramme is the weight of a cubic centimetre of pure water at 4 deg. C. = 0.00220462 lb. 1 erg = 0.00000007373229 foot-lb., and 1 million ergs = 0.07, etc., foot-lb.; 10^7 ergs = 10 million ergs = 0.7, etc., foot-lb. = 1 joule.

ELECTRICAL UNITS—Continued.

Ohm's Law

A law which gives the relations existing in any circuit between current, voltage, and resistance. The formula is: current=voltage÷resistance, which is set down in mathematical form thus: $I=\frac{E}{R}$ (I being the electrical symbol for current, E the symbol for voltage, and R the symbol for resistance). From this equation it is obvious that the voltage can be found by multiplying the current by the resistance ($E=I\times R$), and the resistance is given by dividing the voltage by the current $\left(R=\frac{E}{I}\right)$. In all the above equations the three terms must be in the units of the respective measurements, namely, I in amperes, E in volts, and R in ohms. For an example take a circuit consisting of a battery of 6 volts, across which is joined a resistance of 3 ohms, and this results in a current of 2 amps.

$$\text{Current}=\frac{6}{3}=2\text{ amps.}$$

$$\text{Resistance}=\frac{6}{2}=3\text{ ohms.}$$

$$\text{Voltage}=2\times 3=6\text{ volts.}$$

Ohm's Law for A.C.—Circuits having inductance: $I=\frac{E}{2\pi fL}$.

For A.C. circuits having capacity only, the formula is: $I=2\pi fcE$, or $E=\frac{I}{2\pi fC}$. Where f= frequency in cycles per second, C = capacity, V=voltage, and L=inductance in Henrys. Expressed another way $I=\frac{E}{Z}$, where Z= impedance of circuit.

HEAT

A therm is equal to 100,000 B.Th.U., or $1{\cdot}055 \times 10^5$ ergs.

The Centigrade Heat Unit (C.H.U.) is the heat required to raise 1lb. water 1° Centigrade.

A calorie as used in engineering calculations represents the heat required to raise 1 kilogramme of pure water 1° C. This is the Great Calorie. The Small Calorie represents the heat required to raise 1 gramme of water 1° C.

1 calorie (g.c.) = .0039683 B.Th.U.
1 calorie (e.g.) = 4.1862 joules.
1 calorie (g.c.) = 3.088 foot lbs.
1 calorie (g.c.) = 0.005614 horse-power-second.
1 B.Th.U = 252.00 calories.
1 B.Th.U. = 1005 joules.
1 B.Th.U. = 778.1 foot lbs.
1 B.Th.U. = 1.4147 horse-power-second.
1 C.H.U. = 1.8 B.Th.U.

TIME

1 sidereal second = .99727 second (mean solar).
1 second (mean solar) = 1.002738 sidereal second.
Length of seconds pendulum latitude 45°
= 99.3555 cm.
= 39.1163 inch.

VELOCITY

Velocity of wireless waves = same as light.
Velocity of Earth = 95,000 ft. per sec.
Velocity of light per second = 186,330 miles.
= 299,870 kilometres.
Velocity of sound per second (air 0°C.)
= 1,089 feet
= 331.82 metres.
Velocity of sound per second (brass)
= 11,480 feet
= 3497.96 metres.
Velocity of sound per second (cast steel)
= 16,350 feet
= 4981.85 metres.
Velocity of sound per second (water)
= 4,728 feet
= 1440.62 metres.
Acceleration of falling body = 32 ft per sec./per sec.

PRESSURE

1 pound per square inch

=144 pounds per square foot, which
=0.068 atmosphere
=2.042 inches mercury at 62° F
=27.7 inches water at 62° F.
=2.31 water at 62° F.

1 inch mercury at 62° F.

=1.132 feet of water
=13.58 inches water
=0.491 pounds per square inch

1 atmosphere

= 30 inches mercury at 62° F.
=14.7 pounds per square inch
=2,116.3 pounds per square foot
=33.95 feet of water at 62° F.

1 foot of water at 62° F.

=62.355 pounds per square foot
=0.433 pounds per square inch.

Absolute pressure is the pressure from zero, or pressure of the atmosphere added to indication of pressure gauge.

$$\text{Density} = \frac{\text{Weight of body in air}}{\text{Weight of equal volume of water}}$$

$$\text{Specific Gravity} = \frac{\text{Weight of body in air}}{\text{Weight in air} - \text{Weight in water}}$$

=ratio of Mass of body to volume of water it displaces at 4° C.

EQUIVALENT PRESSURES

Ins. Water	Ins. Mercury	Lbs. per sq. inch	Ins. Water	Ins. Mercury	Lbs. per sq. inch
1″	.074″	.036	19″	1.402″	.684
2″	.148″	.072	20″	1.476″	.720
3″	.221″	.108	21″	1.550″	.756
4″	.295″	.144	22″	1.624″	.792
5″	.369″	.180	23″	1.697″	.828
6″	.443″	.216	24″	1.771″	.864
7″	.517″	.252	25″	1.845″	.900
8″	.590″	.288	26″	1.919″	.936
9″	.664″	.324	27″	1.993″	.972
10″	.738″	.360	28″	2.066″	1.008
11″	.812″	.396	29″	2.140″	1.044
12″	.886″	.432	30″	2.214″	1.080
13″	.959″	.468	31″	2.288″	1.116
14″	1.033″	.504	32″	2.362″	1.152
15″	1.107″	.540	33″	2.435″	1.188
16″	1.180″	.576	34″	2.509″	1.224
17″	1.255″	.612	35″	2.583″	1.260
18″	1.328″	.648	36″	2.657″	1.296

WATER

One cubic foot of water = 62.321 lbs. = .557 cwt. = .028 tons = 6¼ galls.

Head in feet × .4325 = lbs. per square inch (fresh water) R = 2.3122.

Tons water × 224 = gallons ; 1 gallon = 10 lbs.

Cubic feet per minute × 9,000 = gallons per 24 hours.

One atmosphere = 14.7 lbs. per square inch = 33.99 feet water = 29.93 inch mercury.

(*One atmosphere is usually taken* = 15 lbs.)

FORCE

The absolute unit of force is the Poundal, which is that force which, acting for unit time, would impart unit velocity to unit mass.

1 dyne = 0.00007233 poundal.
1 dyne = 0.00102 gram.
1 dyne = 22.48×10^{-7} pounds.
1 megadyne = 1,000,000 dynes.
1 poundal = 13,825 dynes.
1 poundal = 0.03108 pound.
1 poundal = 14.10 grams.

ENERGY

Energy refers to capacity for performing work, or for moving against a resistance.

1 erg = 2.373×10^{-6} foot poundals.
1 erg = $7,376 \times 10^{-8}$ foot pounds.
1 g.cm. = 7.233×10^{-5} foot pounds.
1 joule = 10^{7} ergs.
1 foot poundal = 421,390 ergs.
1 foot pound = 1.35573 joules.
1 foot pound = 13,825-5 g.cm.

The actual energy, Kinetic energy, or dynamic energy of a moving body = $\frac{1}{2}$ mass x velocity^{2}.

POWER

1 watt	= 10^{7} ergs per second.
1 watt	= 23.731 foot poundals per second.
1 watt	= 0.7376 foot lb. per second.
1 watt	= 0.001341 horse-power.
1 kilowatt-hour	= 2,654,200 foot pounds.
1 kilowatt-hour	= 1.3411 horse-power-hour.
1 kilowatt-hour	= 859.975 calories.
1 foot poundal per second	= 421,390 ergs per second.
1 foot pound per second	= 1.35573 watts.
1 horse-power	= 746 watts.
1 horse-power	= 550 foot pounds per second.
1 horse-power	= 178,122 calories per second.

UNITS AND EQUIVALENTS

One ft. lb. ....1 lb. raised 1 foot high.
One BTU......1,055 joules.
One BTU......778.8 ft. lbs.
1 watt ....10^7 ergs per second.
1 watt ....23.731 foot poundals per second.
1 watt ....0.7376 ft. lb. per second.
1 watt ....0.001341 h.p.
One HP hour ..0.746 kW. hour.
One HP hour ..1,980,000 ft. lbs.
One HP hour ..2.545 BTU's.
One kwH. (kilowatt hour) 2,654,200 ft. lbs.
One kwH. ....1,000 watt hours.
One kwH. ....1.34 HP hours.
One kwH. ....3,412 BTU's.
One kwH. ....3,600,000 joules.
One kwH. ....859.975 calories.
One HP ......746 watts.
One HP ......0.746 kW.
One HP ......33,000 ft. lbs. per minute
One HP ......550 ft. lbs. per second.
One HP ......2,545 BTU's per hour.
One HP ......42.4 BTU's per minute.
One HP ......0.707 BTU's per second.
One HP ......178,122 calories per second.

ENGLISH WEIGHTS AND MEASURES

LONG MEASURE

4 inches	=1 hand
12 inches (in.)	=1 foot (ft.)
3 feet	=1 yard (yd.)
5½ yards	=1 rod, pole, or perch
40 poles (220 yards)	=1 furlong (furl.)
8 furlongs (1,760 yards)	=1 mile (m.)
3 miles	=1 league.
1 chain	=100 links (22 yards)
10 chains	=1 furlong
6 feet	=1 fathom
6,080 feet or 1.1516 statute miles	=1 nautical mile

AREA

(Square Measure)

144 square inches	=1 square foot
9 square feet	=1 square yard
30¼ square yards	=1 square pole
40 square poles	=1 rood
4 roods	=1 acre (4,840 sq. yards)
640 acres	=1 square mile

MEASURE OF CAPACITY

(Liquid or Dry Measure)

4 gills	=1 pint	3 bushels	=1 bag
2 pints	=1 quart	4 bushels	=1 coombe
2 quarts	=1 pottle	8 bushels	=1 quarter
2 pottles	=1 gallon	12 bags	=1 chauldron
4 quarts	=1 gallon	5 quarters	=1 load or wey
2 gallons	=1 peck		
4 pecks	=1 bushel	2 loads or weys	=1 last

ENGLISH WEIGHTS AND MEASURES

(*Continued*)

MEASURES OF VOLUME AND CAPACITY

(*Cubic Measure*)

1,728 cubic inches	=1 cubic foot
27 cubic feet	=1 cubic yard
1 marine ton	=40 cubic feet
1 stack	=108 cubic feet
1 cord	=128 cubic feet

Wine Measure

4 gills	=1 pint
2 pints	=1 quart
4 quarts	=1 gallon
10 gallons	=1 anker
18 gallons	=1 runlet or rudlet
31½ gallons	=1 barrel
42 gallons	=1 tierce
63 gallons	=1 hogshead
2 tierces	=1 puncheon
1½ puncheons	=1 pipe or butt
2 pipes	=1 tun

Ale and Beer Measure

4 gills	=1 pint
2 pints	=1 quart
4 quarts	=1 gallon
9 gallons	=1 firkin
2 firkins	=1 kilderkin
2 kilderkins	=1 barrel
1½ barrels	=1 hogshead
1⅓ hogsheads	=1 puncheon
1½ puncheons or 2 hogsheads	=1 butt or pipe
2 pipes	=1 tun

Troy Weight

3.17 grains	=1 carat
24 grains	=1 pennyweight (dwt.)
20 pennyweights	=1 ounce (oz. troy)
12 ounces (troy)	=1 pound (troy)
1 pound (troy)	=5,760 grains
1 pound (avoirdupois)	=7,000 grains (troy)

ENGLISH WEIGHTS AND MEASURES
(Continued)

Avoirdupois Weight

27.34375 grains	=1 dram
16 drams	=1 ounce (oz.)
16 ounces	=1 pound (lb.)
14 pounds	=1 stone
2 stones (28 lbs.)	=1 quarter
100 lbs.	=1 cental
4 quarters	=1 hundredweight (cwt.)
20 cwt.	=1 ton

Apothecaries' Weight

20 grains or minims	=1 scruple
3 scruples	=1 drachm
8 drachms	=1 ounce
12 ounces	=1 pound

Apothecaries' Fluid Measure

60 minims	=1 fluid drachm
8 drachms	=1 fluid ounce
20 ounces	=1 pint (pt. or Octarius)
8 pints	=1 gallon (gal., C., or Congius)

Diamond and Pearl Weight

3.17 grains	=1 carat, or
4 pearl grains	=1 carat
151½ carats	=1 ounce (troy)

Paper Measure

24 sheets	=1 quire
20 quires	=1 ream
2 reams	=1 bundle
10 reams	=1 bale

METRIC SYSTEM

List of Prefixes

mega means a million times.
kilo means a thousand times.
hecto means a hundred times.
deca means ten times.
deci means a tenth part of.
centi means a hundredth part of.
milli means a thousandth part of.
micro means a millionth part of.

Square Measure

100 sq. metres	=1 are.
10,000 sq. metres	=1 hectare.

Weight

10 grammes	=1 decagramme
10 decagrammes	=1 hectogramme.
10 hectogrammes	=1 kilogramme.
1,000 kilogrammes	=1 tonne.

Capacity

1 litre	=1 cubic decimetre.
10 litres	=1 decalitre.
10 decalitres	=1 hectolitre.
10 hectolitres	=1 kilolitre.

Length

10 millimetres	=1 centimetre.
10 centimetres	=1 decimetre.
10 decimetres	=1 metre.
10 metres	=1 decametre.
10 decametres	=1 hectometre.
10 hectometres	=1 kilometre.
10 kilometres	=1 myriametre.

METRIC CONVERSION FACTORS

To convert—

Millimetres to inches .	× .03937 or ÷ 25.4
Centimetres to inches .	× .3937 or ÷ 2.54
Metres to inches . .	× 39.37
Metres to feet . .	× 3.281
Metres to yards . .	× 1.094
Metres per second to feet per minute . .	× 197
Kilometres to miles .	× .6214 or ÷ 1.6093
Kilometres to feet . .	× 3,280.8693
Square millimetres to square inches .	× .00155 or ÷ 645.1
Square centimetres to square inches .	× .155 or ÷ 6.451
Square metres to square feet . . .	× 10.764
Square metres to square yards . . .	× 1.2
Square kilometres to acres	× 247.1
Hectares to acres . .	× 2.471
Cubic centimetres to cubic inches . . .	× .06 or ÷ 16.383
Cubic metres to cubic feet	× 35.315
Cubic metres to cubic yards . . .	× 1.308
Litres to cubic inches .	× 61.022
Litres to gallons . .	× .21998
Hectolitres to cubic feet	× 3.531
Hectolitres to cubic yards	× .131
Grammes to ounces (avoirdupois) . .	× .035 or ÷ 28.35
Grammes per cubic cm. to lb. per cubic inch	÷ 27.7
Joules to foot-lb. . .	× .7373
Kilogrammes to oz. .	× 35.3
Kilogrammes to lb. .	× 2.2046
Kilogrammes to tons .	× .001
Kilogrammes per sq. cm. to lb. per sq. inch. .	× 14.223
Kilogramme - metres to foot-lb. . .	× 7.233

METRIC CONVERSION FACTORS

To convert—

Kilogramme per metre to lb. per foot	× .672
Kilogramme per cubic metre to lb. per cubic foot	× .062
Kilogramme per cheval-vapeur to lb. per h.p.	× 2.235
Kilowatts to h.p.	× 1.34
Watts to h.p.	÷ 746
Watts to foot-lb. per second	× .7373
Cheval-vapeur to h.p.	× .9863
Gallons of water to lb.	× 10
Atmospheres to lb. per sq. inch	× 14.7

Linear Measure Equivalents

1 inch	=2.54 centimetres, or 25.4 millimetres.
1 foot	=30.4799 centimetres, 304.799 millimetres, or .3047 metre.
1 yard	=.914399 metre.
1 mile	=1.6093 kilometres= 5,280 feet.
1 millimetre	=.03937 inch.
1 centimetre	=.3937 inch.
1 decimetre	=3.937 inches.
1 metre	=39.370113 inches. 3.28084 feet. 1.093614 yards.
1 kilometre	=.62137 mile.
1 decametre (10 metres)	=10.936 yards.

Equivalents of Imperial and Metric Weights and Measures

Linear Measure

Imperial		Metric	
1 Inch .	.= 25.400 Millimetres.	1 Millimetre (mm.) (1/1000 m.) .	.= 0.03937 Inch.
1 Foot .	.= 0.30480 Metre.	1 Centimetre (1/100 m.)	= 0.3937 ,,
1 Yard .	.= **0.914399 Metre.**	1 Decimetre (1/10 m.)	= 3.937 Inches.
1 Fathom .	.= 1.8288 Metres.	**1 Metre (m.)** .	.= **39.370113 Inches.** **3.280843 Feet.** **1.0936143 Yards.**
1 Pole .	.= 5.0292 ,,	1 Decametre (10 m.)	= 10.936 Yards
1 Chain .	.= 20.1168 ,,	1 Kilometre (1000 m.)	= 0.62137 Mile
1 Furlong .	.= 201.168 ,,		
1 Mile .	.= 1.6093 Kilometres.		

Square Measure

Imperial		Metric	
1 Square Inch	.= 6.4516 Square Centimetres.	1 Square Centimetre	= 0.15500 Square Inch.
1 Square Foot	.= 9.2903 Square Decimetres.	1 Square Decimetre	= 15.500 Square Inches.
1 Square Yard	.= 0.836126 Square Metre	1 Square Metre	.= 10.7639 Square Feet 1.1960 Square Yards.
1 Rood .	.= 10.117 Ares.	1 Are . .	.= 119.60 Square Yards.
1 Acre .	.= 0.40468 Hectare.	1 Hectare .	.= 2.4711 Acres.
1 Square Mile	.= 259.00 Hectares.		

Equivalents of Imperial and Metric Weights and Measures—continued

Imperial			Metric		
Cubic Measure					
1 Cubic Inch	=	16.387 Cubic Centimetres.	1 Cubic Centimetre	=	0.0610 Cubic Ins.
1 Cubic Foot	=	0.028317 Cubic Metre.	1 Cubic Decimetre (c.d.)	=	61.024 Cubic Ins.
1 Cubic Yard	=	0.764553 ,, ,,	1 Cubic Metre	=	35.3148 Cubic Feet. 1.307954 Cubic Yards.
Measure of Capacity					
1 Pint	=	0.568 Litre.	1 Centilitre (1/100 litre)	=	0.070 Gill.
1 Quart	=	1.136 Litres.	1 Decilitre (1/10 litre)	=	0.176 Pint.
1 Gallon	=	**4.5459631 Litres.**	**1 Litre**	=	**1.75980 Pints.**
Weight					
Avoirdupois					*Avoirdupois*
1 Grain	=	0.0648 Gramme.	1 Milligramme (1/1000 grm.)	=	0.015 Grain.
1 Dram	=	1.772 Grammes.	1 Centigramme (1/100 grm.)	=	0.154 ,,
1 Ounce	=	28.350 ,,	1 Gramme (1 grm.)	=	15.432 ,,
1 Pound (7,000 Grains)	=	**0.45359243 Kilogramme.**	**1 Kilogramme (1,000 grm.)**	=	**2.2046223 Lb. or 15,432.3564 Grains.**
1 Hundredweight	=	50.80 Kilogrammes. 0.5080 Quintal.	1 Quintal (100 kilog.)	=	1.968 cwt.
1 Ton	=	1.0160 Tonnes or 1.016 Kilogrammes.	1 Tonne (1,000 kilog.)	=	0.9842 Ton.
1 Grain (Troy)	=	0.0648 Gramme.	1 Gramme (1 grm.)	=	0.03215 Oz. Troy. 15.432 Grains.
1 Troy Ounce	=	31.1035 Grammes.			

METRIC CONVERSION OF FRACTIONS

($^{1}/_{64}$ inch to 1 inch)

Inches	Dec. Equ.	mm.
1/64	.015625	3969
1/32	.03125	.7937
3/64	.046875	1.1906
1/16	.0625	1.5875
5/64	.078125	1.9844
3/32	.09375	2.3812
7/64	.109375	2.7781
⅛	.125	3.1750
9/64	.140625	3.5719
5/32	.15625	3.9687
11/64	.171875	4.3656
3/16	.1875	4.7625
13/64	.203125	5.1594
7/32	.21875	5.5562
15/64	.234375	5.9531
¼	.25	6.3500
17/64	.265625	6.7469
9/32	.28125	7.1437
19/64	.296875	7.5406
5/16	.3125	7.9375
21/64	.328125	8.3344
11/32	.34375	8.7312
23/64	.359375	9.1281
⅜	.375	9.5250
25/64	.390625	9.9219
13/32	.40625	10.3187
27/64	.421875	10.7156
7/16	.4375	11.1125
29/64	.453125	11.5094
15/32	.46875	11.9062
31/64	.484375	12.3031
½	.5	12.7000
33/64	.515625	13.0969
17/32	.53125	13.4937
35/64	.546875	13.8906
9/16	.5625	14.2875
37/64	.578125	14.6844

METRIC CONVERSION OF FRACTIONS
—*Continued*

($^1/_{64}$ inch to 1 inch)

Inches	Dec. Equ.	mm.
19/32	.59375	15.0812
39/64	.609375	15.4781
⅝	.625	15.8750
41/64	.640625	16.2719
21/32	.65625	16.6687
43/64	.671875	17.0656
11/16	.6875	17.4625
45/64	.703125	17.8594
23/32	.71875	18.2562
47/64	.734375	18.6531
¾	.75	19.0500
49/64	.765625	19.4469
25/32	.78125	19.8437
51/64	.796875	20.2406
13/16	.8125	20.6375
53/64	.828125	21.0344
27/32	.84375	21.4312
55/64	.859375	21.8281
⅞	.875	22.2250
57/64	.890625	22.6219
29/32	.90625	23.0187
59/64	.921875	23.4156
15/16	.9375	23.8125
61/64	.953125	24.2094
31/32	.96875	24.6062
63/64	.984375	25.0031
1	1	25.4000

STANDARD (DENSITY) HEIGHT

Hs=(1.238 Ha+120 To—1800).
Hs=Height in Standard Atmosphere (Density Basis), in feet.
Ha=True Aneroid Height (I.C.A.N. Altimeter), feet.
To=Observed Temperature, °Centigrade.

TABLE OF SLIDE RULE GAUGE POINTS

Known value on Slide	Required value on Rule	Set to On Slide	Set to On Rule
Pounds per square inch	Atmospheres	485	33
Pounds per square inch	Water, head, feet	13	30
Pounds per square inch	Water, head, metres	33	25
Pounds per square inch	Inches, mercury gauge	25	51
Inches, water gauge	Pounds per square inch	360	13
Inches, water gauge	Inches, mercury gauge	14	1
Inches, mercury gauge	Atmospheres	30	1
Atmospheres	Kilos per square centimetre	89	92
Pounds per square foot	Kilos per square metre	87	425
Pounds per lineal foot	Kilos per lineal metre	41	61
Pounds per lineal mile	Kilos per kilometre	71	20
Pounds per cubic foot	Kilos per cubic metre	39	625
Cubic feet of water	Weight in pounds	17	1060
Cubic feet of water	Gallons (imperial)	17	106
Gallons of water	Weight, kilos	108	490
Pounds of water (fresh)	Pounds of water (sea)	38	39
British thermal unit	Calories	250	63
British thermal unit per pound	Calories per kilogramme	9	5
Foot pounds	Kilogrammetres	340	47
Horse-power	Force de cheval	72	73

TABLE OF SLIDE RULE GAUGE POINTS—(*continued*)

Known value on Slide	Required value on Rule	Set to On Slide	Set to On Rule
Pounds per H.P.	Kilos per cheval	300	134
Horse-power per hour	Kilowatts (B.T.U.)	134	100
Watts	Horse-power	5	.0067
Circle, diameter	Circle, circumference	226	710
Circle, diameter	Circle, side of inscribed square	99	70
Circle, diameter	Circle, side of equal square	79	70
Circle, diameter	Circle, side of equal equilateral triangle	72	97
Circle, circumference	Circle, side of inscribed square	40	9
Circle, circumference	Circle, side of equal square	39	11
Circle, area	Inscribed square area	300	191
Square, side	Square, diagonal	70	99
Inches	Centimetres	50	127
Inches, eighths	Millimetres	40	127
Feet	Metres	292	89
Yards	Metres	35	32
Miles	Kilometres	87	140
Square inches	Square centimetres	31	200
Square feet	Square metres	140	13
Square yards	Square metres	161	51

TABLE OF SLIDE RULE GAUGE POINTS—*(continued)*

Known value on Slide	Required value on Rule	Set to On Slide	Set to On Rule
Square miles	Square kilometres	112	290
Acres	Hectares	42	17
Cubic inches	Cubic centimetres	36	590
Cubic feet	Cubic metres	106	3
Cubic feet	Litres	3	85
Cubic yards	Cubic metres	51	39
Gallons	U.S. Gallons	5	6
Bushels	Cubic metres	110	4
Ounces (Avoirdupois)	Grammes	67	1900
Ounces (Avoirdupois)	Kilogrammes	670	19
Pounds (Avoirdupois)	Kilogrammes	280	127
Hundredweights	Kilogrammes	5	254
Tons	Tonnes	62	63
Feet per second	Metres per minute	7	128
Feet per second	Miles per hour	22	15
Feet per minute	Miles per hour	264	3
Yards per minute	Miles per hour	88	3
Miles per hour	Metres per minute	12	322
Knots	Miles per hour	33	38
Pounds per square inch	Kilogrammes per square centimetre	128	9

SPECIFIC GRAVITY AND WEIGHTS

Metals	Weight		Specific Gravity.
	lbs. per cu. in.	lbs. per cu. ft.	
Aluminium, pure cast	.0924	159.63	2.56
Aluminium, pure rolled	.0967	167.11	2.68
Antimony	.2424	418.86	7.71
Brass, cast	.3000	524.00	8.40
Bronze, Tobin ..	.3021	523.06	8.379
Bronze, gun	.3161	546.22	8.750
Bismuth	.3540	611.76	9.80
Cadmium	.3107	536.85	8.60
Copper, pure ..	.3186	550.59	8.82
Copper, rolled ..	.3222	556.83	8.93
Gold, standard ..	.6380	1106.00	17.724
Gold, 24 carat ..	.6979	1206.05	19.32
Iron, cast	.2605	450.08	7.218
Iron, wrought ..	.2779	480.13	7.70
Iron, wire	.2808	485.29	7.774
Iron, pure	.2840	490.66	7.86
Lead	.4108	709.77	11.37
Manganese	.2890	499.40	8.00
Nickel	.3179	549.34	8.80
Platinum	.7767	1342.13	21.50
Platinum, rolled ..	.8280	1436.00	23.00
Steel, Bessemer ..	.2837	479.00	7.852
Steel, soft	.2834	489.74	7.854
Silver	.3805	657.33	10.53
Titanium	.1915	330.85	5.30
Tin	.2634	455.08	7.29
Tungsten	.6900	1192.31	19.10
Uranium	.6755	1167.45	18.70
Vanadium	.1987	343.34	5.30
Zinc, cast	.2479	428.30	6.861
Zinc, pure	.2583	446.43	7.15
Zinc, rolled	.2598	448.90	7.191

Density of Solids and Liquids

Aluminium ..	2.58 g.	per cub cm.	161.1 lb.	per cub. ft.
Copper ..	8.9 g.	,, ,,	555.4 lb.	,, ,,
Gold	19.3 g.	,, ,,	1205 lb.	,, ,,
Ice	0.9167 g.	,, ,,	57.2 lb.	,, ,,
Iron	7.87 g.	,, ,,	491.3 lb.	,, ,,
Lead	11.37 g.	,, ,,	709.77 lb.	,, ,,
Mercury ..	13.596 g.	,, ,,	848.7 lb.	,, ,,
Nickel ..	8.80 g.	,, ,,	549.4 lb.	,, ,,
Platinum ..	21.50 g.	,, ,,	1342.2 lb.	,, ,,
Sea Water ..	1.025 g.	,, ,,	64.0 lb.	,, ,,
Silver ..	10.5 g.	,, ,,	655.5 lb.	,, ,,
Tin	7.18 g.	,, ,,	448.2 lb.	,, ,,
Tungsten ..	18.6 g.	,, ,,	1161.2 lb.	,, ,,
Uranium ..	18.7 g.	,, ,,	1167.4 lb.	,, ,,
Water ..	1.000 g.	,, ,,	62.4 lb.	,, ,,
Zinc	7.19 lb.	,, ,,	448.6 lb.	,, ,,

WEIGHTS OF VARIOUS SUBSTANCES.

Lb. Per Cubic Foot.

LIQUIDS.

Acid, Nitric (91%)	94
„ Sulphuric (87%)	112
Alcohol	49
Benzine	46
Gasoline	42
Mercury	849
Oils..	58
Paraffin	56
Petrol	55
„ refined ..	50
Water, fresh ..	62
„ salt ..	64

METALS.

Aluminium ..	165
Brass	520
Bronze	510
Copper	550
Gold	1205
Gun-metal ..	540
Iron, cast	450
„ wrought ..	480
Lead	710
Nickel	530
Platinum	1342
Silver	655
Steel	490
Tin .. .	460
White-metal ..	460
Zinc	440

SOILS.

Chalk	170
Clay	135
Earth, loose ..	75
Gravel	110
Mud, dry	100
„ wet ..	120
Sand, dry, loose ..	100
„ wet ..	130
Shale	160

STONES, MASONRY.

Brick, pressed ..	150
„ common ..	125
„ soft ..	100
Brickwork ..	112
Cement	90
Concrete . ..	140
„ reinforced	150
„ coke breeze	90
Flint	160
Granite	170
Lime	60
„ mortar .	105
Limestone, comp'r'd	170
„ granular	125
„ loose broken	95
„ walls ..	165

WEIGHTS OF VARIOUS SUBSTANCES.

Lb. per Cubic Foot.

(continued)

Substance	Lb.
Marble	170
Plaster of Paris	140
Rubble masonry	140
Sand, dry, loose	100
Sandstone	150
„ masonry	140
Slate	175
TIMBER.	
Ash	40
Beech	46
Cedar	28
Cherry	36
Chestnut	40
Cork	16
Cypress	30
Ebony	73
Elm	42
„ Canadian	45
Greenheart	70
Hickory	50
Jarrah	57
Larch	38
Mahogany, Spanish	48
„ Honduras	43
Oak, English	59
„ American	59
Pine, White	27
„ Yellow	33
„ Red	34
Pine, Pitch	45
Plane	35
Poplar	26
Spruce	30
Sycamore	40
Teak	50
Walnut	41
MISCELLANEOUS.	
Anthracite, broken, loose	54
Asbestos	187
Asphalt	88
Coal, bituminous	85
„ broken, loose	50
Coke	45
„ loose	30
Flour	40
Glass, window	160
„ flint	190
Grain, Wheat	48
„ Barley	39
„ Oats	32
Hay and Straw, in bales	20
Ice	59
Salt	45
Sulphur	125
White Lead	197

WEIGHTS OF WOODS

The weights of dry woods are as follow:

Substance.	*Weight lb. per cub. ft.*
Alder	33
Almond	43
Ash, American	40
Ash, European	43
Ash, Mountain	43
Balsa	7/8
Bamboo	25
Beech, Common	46
Beech, Australian	33
Birch, American	42
Birch, English	45
Boxwood, Cape	52
Boxwood, West Indian	49
Boxwood, Common	76
Cedar, Cuban	28
Cedar, Virginian	33
Cedar, Indian	28
Cherry, American	36
Cherry, English	38
Chestnut, Sweet	40
Chestnut, Horse	35
Cocus	69
Cogwood	67
Cork	16
Cottonwood, American	34
Cypress	30
Dogwood	49
Ebony	73
Elder	40
Elm, American	44
Elm, Common	42
Fir, Danzig	38
Fir, Riga	36
Fir, Silver	30
Fir, Spruce	30
Hackmatack	39
Hazel	39
Hickory	50
Holly	38
Hornbeam	45
Ironwood	75
Jarrah	57
Juniper	37
Lancewood	57
Larch	38
Lignum-vitæ	83
Lime or Linden	32
Logwood	57
Mahogany, East Indian	43
Mahogany, Cuban	47
Mahogany, Australian	69
Mahogany, Spanish	48
Maple, Bird's-eye	36
Maple, Hard	42
Maple, Soft	38
Oak, African	59
Oak, American	45
Oak, Danzig	52
Oak, English	59
Pine, Pitch	44
Pine, Red	34
Pine, White	27
Pine, Yellow	33
Plane	35
Poplar	26
Rosewood	55
Satinwood	58
Spruce	30
Sycamore	40
Teak	50
Walnut	41
Whitewood	33
Willow	33
Yew	52

Table of Elements

(Arranged according to Mendeleef's Periodic Law)

Atomic Number.	*Element.*	*Atomic Weight.*
First Period :		
1	Hydrogen . .	1.008
2	Helium . .	4.00
Second Period :		
3	Lithium . .	6.94
4	Beryllium . .	9.01
5	Boron . . .	10.82
6	Carbon . .	12.00
7	Nitrogen . .	14.01
8	Oxygen . .	16.00
9	Fluorine . .	19.00
10	Neon . . .	20.2
Third Period :		
11	Sodium . .	23.00
12	Magnesium . .	24.32
13	Aluminium . .	27.1
14	Silicon. . .	28.3
15	Phosphorus . .	31.04
16	Sulphur . .	32.06
17	Chlorine . .	35.46
18	Argon . . .	39.88
Fourth Period :		
19	Potassium . .	39.10
20	Calcium . .	40.07
21	Scandium . .	45.1
22	Titanium . .	48.1
23	Vanadium . .	51.0
24	Chromium . .	52.0
25	Manganese . .	54.93
26	Iron . . .	55.84
27	Cobalt . .	58.97
28	Nickel . . .	58.68

Table of Elements—*Continued*

Atomic Number.	*Element.*	*Atomic Weight.*
Fourth Period:		
29	Copper . .	63.57
30	Zinc . . .	65.37
31	Gallium . .	69.9
32	Germanium . .	72.5
33	Arsenic . .	74.96
34	Selenium . .	79.2
35	Bromine . .	79.92
36	Krypton . .	82.92
Fifth Period:		
37	Rubidium . .	85.45
38	Strontium . .	87.63
39	Yttrium . .	88.7
40	Zirconium . .	90.6
41	Niobium . .	93.5
42	Molybdenum .	96.0
43	Masurium	
44	Ruthenium . .	101.7
45	Rhodium . .	102.9
46	Palladium . .	106.7
47	Silver . . .	107.88
48	Cadmium . .	112.40
49	Indium . .	114.8
50	Tin . . .	118.7
51	Antimony . .	120.2
52	Tellurium . .	127.5
53	Iodine . . .	126.92
54	Xenon . . .	130.2
Sixth Period:		
55	Cæsium . .	132.81
56	Barium . .	137.37
Rare Earths 57	Lanthanum . .	139.0
Rare Earths 58	Cærium . .	140.25
Rare Earths 59	Praseodymium .	140.9

Table of Elements—*Continued*

	Atomic Number.	*Element.*	*Atomic Weight.*
Rare Earths	60	Neodymium	144.3
	61	Illinium	
	62	Samarium	150.4
	63	Europium	152.0
	64	Cadolinium	157.3
	65	Terbium	159.2
	66	Dysprosium	162.5
	67	Holmium	163.5
	68	Erbium	167.7
	69	Thulium	168.5
	70	Ytterbium	173.5
	71	Lutecium	175.0
	72	Hafnium	178.0
	73	Tantalum	181.5
	74	Tungsten	184.0
	75	Rhenium	186.31
	76	Osmium	190.9
	77	Iridium	193.1
	78	Platinum	195.2
	79	Gold	197.2
	80	Mercury	200.6
	81	Thallium	204.4
	82	Lead	207.2
	83	Bismuth	209.0
	84	Polonium	210.0
	85		
	86	Niton	222.0
Seventh Period:			
	87		
	88	Radium	226.0
	89		
	90	Thorium	232.12
	91	Protoactinium	236.0
	92	Uranium	238.2

PROPERTIES OF ELEMENTS

Metal	Symbol	Colour	International Atomic Weight (1934)	Specific Gravity	Specific Heat		Electrical Conductivity Silver at 0° C. = 100	Heat Conductivity Silver = 100	Melting Point		Boiling Point	
					Temp.	Mean			°F.	°C.	°F.	°C.
Aluminium ..	Al	Tin-white ..	26.97	2.70	20—400	0.24	57	35	1218	659	3272	1800
Antimony ..	Sb	Silver-white	121.76	6.62	0—100	0.0495	4	4.02	1166	630	2975 ±15	1635 ±8
Arsenic	As	Steel-grey ..	74.91	5.73	21—268	0.0830	5		1562	850	Sub-limes.	Sub-limes.
Barium	Ba	Yellowish-white	137.36	3.66	—185 to 20	0.0680			1299	704		
Beryllium ..	Be or Gl	Steel-coloured	9.02	1.83	0—100	0.4246			2338	1281		
Bismuth ..		White ..	209.00	9.82	9—102	0.0298	1.3	1.8	514	268	2840	1560
Cadmium ..	Cd	White with blue tinge	112.41	8.64	0—100	0.0548	20	20	610	321	1411	765
Caesium ..	Cs	Silver-white	132.91	1.87	0	0.0522			83	28.25	1238	670
Calcium.. ..	Ca	Yellowish-white	40.08	1.55 (at 29°)	0—100	0.1490	18	25.4	1564	851	2264	1240
Cerium ..	Ce	Steel-grey ..	140.13	6.92	0—100	0.0450			1175	635		
Chromium ..	Cr	Greyish-white	52.01	7.14	22—51	0.1000			3326	1830	3992	2200
Cobalt	Co	Steel-grey ..	58.94	8.79	15—100	0.1030	15	17.2	2673	1467	4379	2415
Columbium (Niobium) ..	Cb Nb	Steel-grey ..	93.3	12.75	0	0.065			3542	1950		

PROPERTIES OF ELEMENTS (*Continued*)

Metal	Symbol	Colour	International Atomic Weight (1934)	Specific Gravity	Specific Heat		Electrical Conductivity Silver at 0° C. = 100	Heat Conductivity Silver = 100	Melting Point		Boiling Point	
					Temp.	Mean			°F.	°C.	°F.	°C.
Copper	Cu	Reddish-yellow	63.57	8.93	15—238	0.0951	94	92	1981	1082.6	4190	2310
Dysprosium ..	Dy		162.46									
Erbium ..	Er		167.64	4.77(?)								
Europium ..	Eu		152.0									
Gadolinium ..	Gd		157.3									
Gallium.. ..	Ga	Silver-white	69.72	5.95	0	0.079			86	30	3632	2000
Germanium ..	Ge	Greyish-white	72.60	5.47	0	0.0737			1652	900		
Gold	Au	Yellow ..	197.2	19.32	0	0.0316	67	70	1945.5	1063	3992	2200
Hafnium ..	Hf		178.6	13.08								
Indium	In	Silver-white	114.76	7.12	0	0.0570			311	155		
Iridium	Ir	Grey ..	193.1	22.41	0—100	0.0323			4451	2454		
Iron	Fe	Greyish-white	55.84	7.87	20—100	0.1190	17	16	2780.6 ±5	1527 ±3	4442	2450
Lanthanum ..	La	White ..	138.9	6.12	0	0.0449			1490	810		
Lead	Pb	Blue-grey ..	207.22	11.37	18—100	0.0310	7.2	8.5	621	327	3132	1740
Lithium.. ..	Li	Silver-white	6.94	0.534	0—100	0.9600	16		367	186	2437	1336
Magnesium ..	Mg	Silver-white	24.32	1.74	18—99	0.2460	34	34.3	1200	649	2048	1120
Manganese ..	Mn	White-grey	54.93	7.39	0	0.1217			2268	1242	3452	1900
Mercury.. ..	Hg	White ..	200.61	13.56 (at 15° C)	20—50	0.0331	1.5	5.3	—38	—38.5	673	356.7

PROPERTIES OF ELEMENTS (*Continued*)

Metal	Symbol	Colour	International Atomic Weight (1934)	Specific Gravity	Specific Heat		Electrical Conductivity Silver at 0° C. = 100	Heat Conductivity Silver = 100	Melting Point		Boiling Point	
					Temp.	Mean			°F.	°C.	°F.	°C.
Molybdenum ..	Mc	Dull silver .	96.0	10.0	15—91	0.0723			4532	2500		
Neodymium ..	Nd		144.27	6.96					1544	840		
Nickel	Ni	White ..	58.69	8.90	0—100	0.1147	20.5	14	2651	1455	4244	2340
Osmium ..	Os	Blue-white	191.5	22.48	18—98	0.0311			4532	2500		
Palladium ..	Pd	White ..	106.7	12.16	18—100	0.059			2831	1555	3992	2200
Platinum ..	Pt	White ..	195.23	21.4	0—100	0.0323	13.5	37.9	3224	1773.5	7772	4300
Potassium ..	K	Silver-white	39.096	0.862	0	0.1728	17	45	144	62.5	1404	762.2
Praseodymium ..	Pr	White ..	140.92	6.48	0	0.0314			1724	940		
Radium.. ..	Ra	Brilliant-white	225.97						1292	700		
Rhenium ..	Re		186.31	21.4					6224 ±108	3440 ±60		
Rhodium ..	Rh	Bluish-white	102.91	12.41	0	0.0580			3571 ±5	1966 ±3		
Rubidium ..	Rb	White ..	85.44	1.532	0	0.0802			100	38	1285	696
Ruthenium ..	Ru	White ..	101.7	12.3	0	0.0611						
Samarium ..	Sm		150.43	7.8					2462	1350		
Scandium ..	Sc		45.10									
Selenium ..	Se	Steel-grey ..	78.96	4.5	15—217	0.084	Varies		423	217	1274	690
Silver	Ag	White ..	107.88	10.5	17—517	0.059	100	100	1760	960	3550	1955
Sodium	Na	Silver-white	22.997	0.971	0	0.2811	28	36.5	207	97.5	1622	882.9

PROPERTIES OF ELEMENTS (*Continued*)

Metal	Symbol	Colour	International Atomic Weight (1934)	Specific Gravity	Specific Heat		Electrical Conductivity Silver at 0° C. = 100	Heat Conductivity Silver = 100	Melting Point		Boiling Point	
					Temp.	Mean			°F.	°C.	°F.	°C.
Strontium ..	Sr	Yellowish-white	87.63	2.74	0	0.0735			1420	771		
Tantalum ..	Ta	Iron-grey ..	181.4	16.6	—185 to 20	0.0326			5270	2910		
Tellurium ..	Te	Shining-white	127.61	6.25	15—100	0.0483	0.077		842	450	2534	1390
Terbium ..	Tr		159.2									
Thallium ..	Tl	White ..	204.39	11.9	20—100	0.0326	8		574	301	2654	1457
Thorium ..	Th	Greyish-white	232.12	11.3	0	0.0276			3090	1700		
Thulium ..	Tm		169.4									
Tin	Sn	Silver-white	118.70	7.29	0—100	0.0559	11.3	15.2	450	232	4118	2270
Titanium ..	Ti	Dark-grey ..	47.90	4.8	0—100	0.1125			3632	2000		
Tungsten ..	W	Steel-grey ..	184.0	19.2	15—93	0.0340			6107	3375		
Uranium ..	U	Silver-white	238.14	18.7	0	0.028			3360	1800		
Vanadium ..	V	Light-grey .	50.95	5.5	0—100	0.1153			3128	1720		
Ytterbium ..	Yb		173.04									
Yttrium ..	Y	Dark-grey ..	88.92	3.8(?)								
Zinc	Zn	Bluish-white	65.38	7.1	20—100	0.0931	25.5	26.3	786.9	419.4	1662	905.7
Zirconium ..	Zr	Grey ..	91.22	6.53	0	0.0662			3501	1927		

(*These tables are by courtesy of "Metal Industry."*)

COMPARISON OF THERMOMETERS

The thermometer scales most generally in use are the Fahrenheit (F.) which is used in all English-speaking countries; the Centigrade or Celsius (C.) which is used in most of the Continental countries and also in scientific work; and the Reaumur (R.) which is also used to some extent in this country and also on the Continent.

The freezing point of water is marked at 32° on the Fahrenheit scale and the boiling point at 212° at atmospheric pressure. The distance between freezing point and boiling point is divided into 180°.

With the Centigrade scale the freezing point of water is 0° and the boiling point 100°.

On the Reaumur scale the freezing point is at 0° and the boiling point at 80°.

For converting temperatures on any one of the three scales to the other the following formulæ apply:

$$\text{Degrees Fahrenheit} = \frac{9 \times \text{degrees C.}}{5} + 32$$

$$= \frac{9 \times \text{degrees R.}}{4} + 32$$

$$\text{Degrees Centigrade} = \frac{5 \times (\text{degrees F.} - 32)}{9}$$

$$= \frac{5 \times \text{degrees R.}}{4}$$

$$\text{Degrees Reaumur} = \frac{4 \times \text{degrees C.}}{5}$$

$$= \frac{4 \times (\text{degrees F.} - 32)}{9}$$

Absolute zero refers to that point on the thermometer scale (theoretically determined) where a lower temperature is inconceivable. This is located at − 273° C. or − 459.2° F., and any temperature reckoned from absolute zero instead of from the normal zero is termed absolute pressure. Hence to find absolute temperature from the temperature expressed in degrees F. add 459.2 to the degrees F.

TEMPERATURE CONVERSION TABLE

NOTE.—Find the known temperature to be converted in the column marked A. Then read the Centigrade conversion to left and Fahrenheit to right.

e.g.: −2.8 27 80.6
27°C. = 80.6°F.
27°F = −2.8°C.

°C.	A	°F.	°C.	A	°F.
−273	−459		−68	−90	−130
−268	−450		−62	−80	−112
−262	−440		−57	−70	−94
−257	−430		−51	−60	−76
−251	−420		−45.6	−50	−58.0
−246	−410		−45.0	−49	−56.2
−240	−400		−44.4	−48	−54.4
−234	−390		−43.9	−47	−52.6
−229	−380		−43.3	−46	−50.8
−223	−370		−42.8	−45	−49.0
−218	−360		−42.2	−44	−47.2
−212	−350		−41.7	−43	−45.4
−207	−340		−41.1	−42	−43.6
−201	−330		−40.6	−41	−41.8
−196	−320		−40.0	−40	−40.0
−190	−310		−39.4	−39	−38.2
−184	−300		−38.9	−38	−36.4
−179	−290		−38.3	−37	−34.6
−173	−280		−37.8	−36	−32.8
−169	−273	−459.4	−37.2	−35	−31.0
−168	−270	−454	−36.7	−34	−29.2
−162	−260	−436	−36.1	−33	−27.4
−157	−250	−418	−35.6	−32	−25.6
−151	−240	−400	−35.0	−31	−23.8
−146	−230	−382	−34.4	−30	−22.0
−140	−220	−364	−33.9	−29	−20.2
−134	−210	−346	−33.3	−28	−18.4
−129	−200	−328	−32.8	−27	−16.6
−123	−190	−310	−32.2	−26	−14.8
−118	−180	−292	−31.7	−25	−13.0
−112	−170	−274	−31.1	−24	−11.2
−107	−160	−256	−30.6	−23	−9.4
−101	−150	−238	−30.0	−22	−7.6
−96	−140	−220	−29.4	−21	−5.8
−90	−130	−202	−28.9	−20	−4.0
−84	−120	−184	−28.3	−19	−2.2
−79	−110	−166	−27.8	−18	−0.4
−73	−100	−148	−27.2	−17	1.4

TEMPERATURE CONVERSION TABLE
(Continued)

°C.	A	°F.	°C.	A	F.
−26.7	−16	3.2	−3.3	26	78.8
−26.1	−15	5.0	−2.8	27	80.6
−25.6	−14	6.8	−2.2	28	82.4
−25.0	−13	8.6	−1.7	29	84.2
−24.4	−12	10.4	−1.1	30	86.0
−23.9	−11	12.2	−0.6	31	87.8
−23.3	−10	14.0	0.	32	89.6
−22.8	−9	15.8	0.6	33	91.4
−22.2	−8	17.6	1.1	34	93.2
−21.7	−7	19.4	1.7	35	95.0
−21.1	−6	21.2	2.2	36	96.8
−20.6	−5	23.0	2.8	37	98.6
−20.0	−4	24.8	3.3	38	100.4
−19.4	−3	26.6	3.9	39	102.2
−18.9	−2	28.4	4.4	40	104.0
−18.3	−1	30.2	5.0	41	105.8
−17.8	0	32.0	5.6	42	107.6
−17.2	1	33.8	6.1	43	109.4
−16.7	2	35.6	6.7	44	111.2
−16.1	3	37.4	7.2	45	113.0
−15.6	4	39.2	7.8	46	114.8
−15.0	5	41.0	8.3	47	116.6
−14.4	6	42.8	8.9	48	118.4
−13.9	7	44.6	9.4	49	120.2
−13.3	8	46.4	10.0	50	122.0
−12.8	9	48.2	10.6	51	123.8
−12.2	10	50.0	11.1	52	125.6
−11.7	11	51.8	11.7	53	127.4
−11.1	12	53.6	12.2	54	129.2
−10.6	13	55.4	12.8	55	131.0
−10.0	14	57.2	13.3	56	132.8
−9.4	15	59.0	13.9	57	134.6
−8.9	16	60.8	14.4	58	136.4
−8.3	17	62.6	15.0	59	138.2
−7.8	18	64.4	15.6	60	140.0
−7.2	19	66.2	16.1	61	141.8
−6.7	20	68.0	16.7	62	143.6
−6.1	21	69.8	17.2	63	145.4
−5.6	22	71.6	17.8	64	147.2
−5.0	23	73.4	18.3	65	149.0
−4.4	24	75.2	18.9	66	150.8
−3.9	25	77.0	19.4	67	152.6

TEMPERATURE CONVERSION TABLE

(Continued)

°C.	A	°F.	°C.	A	°F.
20.0	68	154.4	43.3	110	230.0
20.6	69	156.2	43.9	111	231.8
21.1	70	158.0	44.4	112	233.6
21.7	71	159.8	45.0	113	235.4
22.2	72	161.6	45.6	114	237.2
22.8	73	163.4	46.1	115	239.0
23.3	74	165.2	46.7	116	240.8
23.9	75	167.0	47.2	117	242.6
24.4	76	168.8	47.8	118	244.4
25.0	77	170.6	48.3	119	246.2
25.6	78	172.4	48.9	120	248.0
26.1	79	174.2	49.4	121	249.8
26.7	80	176.0	50.0	122	251.6
27.2	81	177.8	50.6	123	253.4
27.8	82	179.6	51.1	124	255.2
28.3	83	181.4	51.7	125	257.0
28.9	84	183.2	52.2	126	258.8
29.4	85	185.0	52.8	127	260.6
30.0	86	186.8	53.3	128	262.4
30.6	87	188.6	53.9	129	264.2
31.1	88	190.4	54.4	130	266.0
31.7	89	192.2	55.0	131	267.8
32.2	90	194.0	55.6	132	269.6
32.8	91	195.8	56.1	133	271.4
33.3	92	197.6	56.7	134	273.2
33.9	93	199.4	57.2	135	275.0
34.4	94	201.2	57.8	136	276.8
35.0	95	203.0	58.3	137	278.6
35.6	96	204.8	58.9	138	280.4
36.1	97	206.6	59.4	139	282.2
36.7	98	208.4	60.0	140	284.0
37.2	99	210.2	60.6	141	285.8
37.8	100	212.0	61.1	142	287.6
38.3	101	213.8	61.7	143	289.4
38.9	102	215.6	62.2	144	291.2
39.4	103	217.4	62.8	145	293.0
40.0	104	219.2	63.3	146	294.8
40.6	105	221.0	63.9	147	296.6
41.1	106	222.8	64.4	148	298.4
41.7	107	224.6	65.0	149	300.2
42.2	108	226.4	65.6	150	302.0
42.8	109	228.2	66.1	151	303.8

TEMPERATURE CONVERSION TABLE
(*Continued*)

°C.	A	°F.	°C.	A	°F.
66.7	152	305.6	90.0	194	381.2
67.2	153	307.4	90.6	195	383.0
67.8	154	309.2	91.1	196	384.8
68.3	155	311.0	91.7	197	386.6
68.9	156	312.8	92.2	198	388.4
69.4	157	314.6	92.8	199	390.2
70.0	158	316.4	93.3	200	392.0
70.6	159	318.2	93.9	201	393.8
71.1	160	320.0	94.4	202	395.6
71.7	161	321.8	95.0	203	397.4
72.2	162	323.6	95.6	204	399.2
72.8	163	325.4	96.1	205	401.0
73.3	164	327.2	96.7	206	402.8
73.9	165	329.0	97.2	207	404.6
74.4	166	330.8	97.8	208	406.4
75.0	167	332.6	98.3	209	408.2
75.6	168	334.4	98.9	210	410.0
76.1	169	336.2	99.4	211	411.8
76.7	170	338.0	100.0	212	413.6
77.2	171	339.8	100.6	213	415.4
77.8	172	341.6	101.1	214	417.2
78.3	173	343.4	101.7	215	419.0
78.9	174	345.2	102.2	216	420.8
79.4	175	347.0	102.8	217	422.6
80.0	176	348.8	103.3	218	424.4
80.6	177	350.6	103.9	219	426.2
81.1	178	352.4	104.4	220	428.0
81.7	179	354.2	105.0	221	429.8
82.2	180	356.0	105.6	222	431.6
82.8	181	357.8	106.1	223	433.4
83.3	182	359.6	106.7	224	435.2
83.9	183	361.4	107.2	225	437.0
84.4	184	363.2	107.8	226	438.8
85.0	185	365.0	108.3	227	440.6
85.6	186	366.8	108.9	228	442.4
86.1	187	368.6	109.4	229	444.2
86.7	188	370.4	110.0	230	446.0
87.2	189	372.2	110.6	231	447.8
87.8	190	374.0	111.1	232	449.6
88.3	191	375.8	111.7	233	451.4
88.9	192	377.6	112.2	234	453.2
89.4	193	379.4	112.8	235	455.0

TEMPERATURE CONVERSION TABLE
(*Continued*)

°C.	A	°F.	°C.	A	°F.
113.3	236	456.8	277	530	986
113.9	237	458.6	282	540	1004
114.4	238	460.4	288	550	1022
115.0	239	462.2	293	560	1040
115.6	240	464.0	299	570	1058
116.1	241	465.8	304	580	1076
116.7	242	467.6	310	590	1094
117.2	243	469.4	316	600	1112
117.8	244	471.2	321	610	1130
118.3	245	473.0	327	620	1148
118.9	246	474.8	332	630	1166
119.4	247	476.6	338	640	1184
120.0	248	478.4	343	650	1202
120.6	249	480.2	349	660	1220
121	250	482	354	670	1238
127	260	500	360	680	1256
132	270	518	366	690	1274
138	280	536	371	700	1292
143	290	554	377	710	1310
149	300	572	382	720	1328
154	310	590	388	730	1346
160	320	608	393	740	1364
166	330	626	399	750	1382
171	340	644	404	760	1400
177	350	662	410	770	1418
182	360	680	416	780	1436
188	370	698	421	790	1454
193	380	716	427	800	1472
199	390	734	432	810	1490
204	400	752	438	820	1508
210	410	770	443	830	1526
216	420	788	449	840	1544
221	430	806	454	850	1562
227	440	824	460	860	1580
232	450	842	466	870	1598
238	460	860	471	880	1616
243	470	878	477	890	1634
249	480	896	482	900	1652
254	490	914	488	910	1670
260	500	932	493	920	1688
266	510	950	499	930	1706
271	520	968	504	940	1724

TEMPERATURE CONVERSION TABLE
(*Continued*)

°C.	A	F.	°C.	A	°F.
510	950	1742	666	1230	2246
516	960	1760	671	1240	2264
521	970	1778	677	1250	2282
527	980	1796	682	1260	2300
532	990	1814	688	1270	2318
538	1000	1832	693	1280	2336
543	1010	1850	699	1290	2354
549	1020	1868	704	1300	2372
554	1030	1886	710	1310	2390
560	1040	1904	716	1320	2408
566	1050	1922	721	1330	2426
571	1060	1940	727	1340	2444
577	1070	1958	732	1350	2462
582	1080	1976	738	1360	2480
588	1090	1994	743	1370	2498
593	1100	2012	749	1380	2516
599	1110	2030	754	1390	2534
604	1120	2048	760	1400	2552
610	1130	2066	766	1410	2570
616	1140	2084	771	1420	2588
621	1150	2102	777	1430	2606
627	1160	2120	782	1440	2624
632	1170	2138	788	1450	2642
638	1180	2156	793	1460	2660
643	1190	2174	799	1470	2678
649	1200	2192	804	1480	2696
654	1210	2210	810	1490	2714
660	1220	2228	816	1500	2732

VALUES OF SINGLE DEGREES

°C.	°F.	°F.	°C.
1 =	1.8	1 =	0.56
2 =	3.6	2 =	1.11
3 =	5.4	3 =	1.67
4 =	7.2	4 =	2.22
5 =	9.0	5 =	2.78
6 =	10.8	6 =	3.33
7 =	12.6	7 =	3.89
8 =	14.4	8 =	4.44
9 =	16.2	9 =	5.0

$°C. = 5/9\,(°F. - 32°)$ $\qquad$ $°F. = (9/5\,°C.) + 32°$

CHORDS AND RADIANS

Degrees	Chords	Differences for 10′	Radians	Constant Differences	
0°	.0000	29	.0000	1′	3″
1°	.0175	29	.0175	2′	6″
2°	.0349	29	.0349	3′	9″
3°	.0524	29	.0524	4′	12″
4°	.0698	29	.0698	5′	15″
5°	.0872	29	.0873	6′	17″
6°	.1047	29	.1047	7′	20″
7°	.1221	29	.1222	8′	23″
8°	.1395	29	.1396	9′	26″
9°	.1569	29	.1571	10′	29″
10°	.1743	29	.1745		—
11°	.1917	29	.1920		
12°	.2091	29	.2094		
13°	.2264	29	.2269		
14°	.2437	29	.2443		
15°	.2611	29	.2618		
16°	.2783	29	.2793		
17°	.2956	29	.2967		
18°	.3129	29	.3142		
19°	.3301	29	.3316		
20°	.3473	29	.3491		
21°	.3645	29	.3665		
22°	.3816	29	.3840		
23°	.3987	28	.4014		
24°	.4153	28	.4189		
25°	.4329	28	.4363		
26°	.4499	28	.4538		
27°	.4669	28	.4712		
28°	.4833	28	.4887		
29°	.5008	28	.5061		
30°	.5176	28	.5236		
31°	.5345	28	.5411		
32°	.5513	28	.5585		
33°	.5680	28	.5760		
34°	.5847	28	.5934		
35°	.6014	28	.6109		
36°	.6180	28	.6283		
37°	.6346	28	.6458		
38°	.6511	27	.6632		
39°	.6676	27	.6807		
40°	.6840	27	.6981		
41°	.7004	27	.7156		
42°	.7167	27	.7330		
43°	.7330	27	.7505		
44°	.7492	27	.7679		
45°	.7654	27	.7854		

1 right angle $= \frac{\pi}{2}$ radians = 1.5707963 radians

log. 1.57079 = .1961

1 radian = 57.2958° = 57° 17′ 45″

log. 57.2958 = 1.7581

To convert radians into seconds multiply by 206265.

log. 206265 = 5.3144.

CHORDS AND RADIANS (*Contd.*)

Degrees	Chords	Differences for 10′	Radians	Constant Differences
45°	.7654	27	.7854	1′ 3″
46°	.7815	27	.8029	2′ 6″
47°	.7975	27	.8203	3′ 9″
48°	.8135	27	.8378	4′ 12″
49°	.8294	26	.8552	5′ 15″
50°	.8452	26	.8727	6′ 17″
51°	.8610	26	.8901	7′ 20″
52°	.8767	26	.9076	8′ 23″
53°	.8924	26	.9250	9′ 26″
54°	.9080	26	.9425	10′ 29″
55°	.9235	26	.9599	—
56°	.9389	26	.9774	
57°	.9543	26	.9948	
58°	.9696	25	1.0123	
59°	.9848	25	1.0297	
60°	1.0000	25	1.0472	
61°	1.0151	25	1.0647	
62°	1.0301	25	1.0821	
63°	1.0450	25	1.0996	
64°	1.0598	25	1.1170	
65°	1.0746	24	1.1345	
66°	1.0893	24	1.1519	
67°	1.1039	24	1.1694	
68°	1.1184	24	1.1868	
69°	1.1328	24	1,2043	
70°	1.1472	24	1.2217	
71°	1.1614	24	1.2392	
72°	1.1756	23	1.2566	
73°	1.1896	23	1.2741	
74°	1.2036	23	1.2915	
75°	1.2175	23	1,3090	
76°	1.2313	23	1,3265	
77°	1.2450	23	1.3439	
78°	1.2586	23	1.3614	
79°	1.2722	22	1.3788	
80°	1.2856	22	1.3963	
81°	1.2989	22	1.4137	
82°	1.3121	22	1.4312	
83°	1.3252	22	1.4486	
84°	1.3383	22	1.4661	
85°	1.3512	21	1.4835	
86°	1.3640	21	1.5010	
87°	1.3767	21	1.5184	
88°	1.3893	21	1.5359	
89°	1.4018	21	1.5533	
90°	1.4142	—	1.5708	

B. Chord = diameter × sine of angle subtended at circumference.
= diameter × sine of semi-angle subtended at centre.

TABLE FOR CONVERTING MINUTES INTO DECIMALS OF A DEGREE

Min.	Dec. of Degree	Min.	Dec. of Degree	Min.	Dec. of Degree	Min.	Dec. of Degree	Min.	Dec. of Degree
¼	0.00416	12¼	0.20416	24¼	0.40416	36¼	0.60416	48¼	0.80416
½	0.00833	12½	0.20833	24½	0.40833	36½	0.60833	48½	0.80833
¾	0.01250	12¾	0.21250	24¾	0.41250	36¾	0.61250	48¾	0.81250
1	0.01666	13	0.21666	25	0.41666	37	0.61666	49	0.81666
1¼	0.02083	13¼	0.22083	25¼	0.42083	37¼	0.62083	49¼	0.82083
1½	0.02500	13½	0.22500	25½	0.42500	37½	0.62500	49½	0.82500
1¾	0.02916	13¾	0.22916	25¾	0.42916	37¾	0.62916	49¾	0.82916
2	0.03333	14	0.23333	26	0.43333	38	0.63333	50	0.83333
2¼	0.03750	14¼	0.23750	26¼	0.43750	38¼	0.63750	50¼	0.83750
2½	0.04166	14½	0.24166	26½	0.44166	38½	0.64166	50½	0.84166
2¾	0.04583	14¾	0.24583	26¾	0.44583	38¾	0.64583	50¾	0.84583
3	0.05000	15	0.25000	27	0.45000	39	0.65000	51	0.85000
3¼	0.05416	15¼	0.25416	27¼	0.45416	39¼	0.65416	51¼	0.85416
3½	0.05833	15½	0.25833	27½	0.45833	39½	0.65833	51½	0.85833
3¾	0.06250	15¾	0.26250	27¾	0.46250	39¾	0.66250	51¾	0.86250
4	0.06666	16	0.26666	28	0.46666	40	0.66666	52	0.86666

From Foote Bros., by permission.

TABLE FOR CONVERTING MINUTES INTO DECIMALS OF A DEGREE—(*Continued*)

Min.	Dec. of Degree	Min.	Dec. of Degree	Min.	Dec. of Degree	Min.	Dec. of Degree	Min.	Dec. of Degree
4¼	0.07083	16¼	0.27083	28¼	0.47083	40¼	0.67083	52¼	0.87083
4½	0.07500	16½	0.27500	28½	0.47500	40½	0.67500	52½	0.87500
4¾	0.07916	16¾	0.27916	28¾	0.47916	40¾	0.67916	52¾	0.87916
5	0.08333	17	0.28333	29	0.48333	41	0.68333	53	0.88333
5¼	0.08750	17¼	0.28750	29¼	0.48750	41¼	0.68750	53¼	0.88750
5½	0.09166	17½	0.29166	29½	0.49166	41½	0.69166	53½	0.89166
5¾	0.09583	17¾	0.29583	29¾	0.49583	41¾	0.69583	53¾	0.89583
6	0.10000	18	0.30000	30	0.50000	42	0.70000	54	0.90000
6¼	0.10416	18¼	0.30416	30¼	0.50416	42¼	0.70416	54¼	0.90416
6½	0.10833	18½	0.30833	30½	0.50833	42½	0.70833	54½	0.90833
6¾	0.11250	18¾	0.31250	30¾	0.51250	42¾	0.71250	54¾	0.91250
7	0.11666	19	0.31666	31	0.51666	43	0.71666	55	0.91666
7¼	0.12083	19¼	0.32083	31¼	0.52083	43¼	0.72083	55¼	0.92083
7½	0.12500	19½	0.32500	31½	0.52500	43½	0.72500	55½	0.92500
7¾	0.12916	19¾	0.32916	31¾	0.52916	43¾	0.72916	55¾	0.92916
8	0.13333	20	0.33333	32	0.53333	44	0.73333	56	0.93333

From Foote Bros., by permissi on.

TABLE FOR CONVERTING MINUTES INTO DECIMALS OF A DEGREE—(*Continued*)

Min.	Dec. of Degree	Min.	Dec. of Degree	Min.	Dec. of Degree	Min.	Dec. of Degree	Min.	Dec. of Degree
8¼	0.13750	20¼	0.33750	32¼	0.53750	44¼	0.73750	56¼	0.93750
8½	0.14166	20½	0.34166	32½	0.54166	44½	0.74166	56½	0.94166
8¾	0.14583	20¾	0.34583	32¾	0.54583	44¾	0.74583	56¾	0.94583
9	0.15000	21	0.35000	33	0.55000	45	0.75000	57	0.95000
9¼	0.15416	21¼	0.35416	33¼	0.55416	45¼	0.75416	57¼	0.95416
9½	0.15833	21½	0.35833	33½	0.55833	45½	0.75833	57½	0.95833
9¾	0.16250	21¾	0.36250	33¾	0.56250	45¾	0.76250	57¾	0.96250
10	0.16666	22	0.36666	34	0.56666	46	0.76666	58	0.96666
10¼	0.17083	22¼	0.37083	34¼	0.57083	46¼	0.77083	58¼	0.96083
10½	0.17500	22½	0.37500	34½	0.57500	46½	0.77500	58½	0.97500
10¾	0.17916	22¾	0.37916	34¾	0.57916	46¾	0.77916	58¾	0.97916
11	0.18333	23	0.38333	35	0.58333	47	0.78333	59	0.98333
11¼	0.18750	23¼	0.38750	35¼	0.58750	47¼	0.78750	59¼	0.98750
11½	0.19166	23½	0.39166	35½	0.59166	47½	0.79166	59½	0.99166
11¾	0.19583	23¾	0.39583	35¾	0.59583	47¾	0.79583	59¾	0.99583
12	0.20000	24	0.40000	36	0.60000	48	0.80000	60	1.00000

From Foote Bros., by permission.

NATURAL SINES

Degrees	0′	6′	12′	18′	24′	30′	36′	42′	48′	54′	Mean Differences 1′	2′	3′	4′	5′
0	.0000	0017	0035	0052	0070	0087	0105	0122	0140	0157	3	6	9	12	15
1	.0175	0192	0209	0227	0244	0262	0279	0297	0314	0332	3	6	9	12	15
2	.0349	0366	0384	0401	0419	0436	0454	0471	0488	0506	3	6	9	12	15
3	.0523	0541	0558	0576	0593	0610	0628	0645	0663	0680	3	6	9	12	15
4	.0698	0715	0732	0750	0767	0785	0802	0819	0837	0854	3	6	9	12	14
5	.0872	0889	0906	0924	0941	0958	0976	0993	1011	1028	3	6	9	12	14
6	.1045	1063	1080	1097	1115	1132	1149	1167	1184	1201	3	6	9	12	14
7	.1219	1236	1253	1271	1288	1305	1323	1340	1357	1374	3	6	9	12	14
8	.1392	1409	1426	1444	1461	1478	1495	1513	1530	1547	3	6	9	12	14
9	.1564	1582	1599	1616	1633	1650	1668	1685	1702	1719	3	6	9	12	14
10	.1736	1754	1771	1788	1805	1822	1840	1857	1874	1891	3	6	9	11	14
11	.1908	1925	1942	1959	1977	1994	2011	2028	2045	2062	3	6	9	11	14
12	.2079	2096	2113	2130	2147	2164	2181	2198	2215	2233	3	6	9	11	14
13	.2250	2267	2284	2300	2317	2334	2351	2368	2385	2402	3	6	8	11	14
14	.2419	2436	2453	2470	2487	2504	2521	2538	2554	2571	3	6	8	11	14
15	.2588	2605	2622	2639	2656	2672	2689	2706	2723	2740	3	6	8	11	14
16	.2756	2773	2790	2807	2823	2840	2857	2874	2890	2907	3	6	8	11	14
17	.2924	2940	2957	2974	2990	3007	3024	3040	3057	3074	3	6	8	11	14

NATURAL SINES (Continued)

Degrees	0′	6′	12′	18′	24′	30′	36′	42′	48′	54′	Mean Differences				
											1′	2′	3′	4′	5′
18	.3090	3107	3123	3140	3156	3173	3190	3206	3223	3239	3	6	8	11	14
19	.3256	3272	3289	3305	3322	3338	3355	3371	3387	3404	3	5	8	11	14
20	.3420	3437	3453	3469	3486	3502	3518	3535	3551	3567	3	5	8	11	14
21	.3584	3600	3616	3633	3649	3665	3681	3697	3714	3730	3	5	8	11	14
22	.3746	3762	3778	3795	3811	3827	3843	3859	3875	3891	3	5	8	11	14
23	.3907	3923	3939	3955	3971	3987	4003	4019	4035	4051	3	5	8	11	14
24	.4067	4083	4099	4115	4131	4147	4163	4179	4195	4210	3	5	8	11	13
25	.4226	4242	4258	4274	4289	4305	4321	4337	4352	4368	3	5	8	11	13
26	.4384	4399	4415	4431	4446	4462	4478	4493	4509	4524	3	5	8	10	13
27	.4540	4555	4571	4586	4602	4617	4633	4648	4664	4679	3	5	8	10	13
28	.4695	4710	4726	4741	4756	4772	4787	4802	4818	4833	3	5	8	10	13
29	.4848	4863	4879	4894	4909	4924	4939	4955	4970	4985	3	5	8	10	13
30	.5000	5015	5030	5045	5060	5075	5090	5105	5120	5135	3	5	8	10	13
31	.5150	5165	5180	5195	5210	5225	5240	5255	5270	5284	2	5	7	10	12
32	.5299	5314	5329	5344	5358	5373	5388	5402	5417	5432	2	5	7	10	12
33	.5446	5461	5476	5490	5505	5519	5534	5548	5563	5577	2	5	7	10	12
34	.5592	5606	5621	5635	5650	5664	5678	5693	5707	5721	2	5	7	10	12
35	.5736	5750	5764	5779	5793	5807	5821	5835	5850	5864	2	5	7	9	12

NATURAL SINES (Continued)

Degrees	0′	6′	12′	18′	24′	30′	36′	42′	48′	54′	Mean Differences 1′	2′	3′	4′	5′
36	.5878	5892	5906	5920	5934	5948	5962	5976	5990	6004	2	5	7	9	12
37	.6018	6032	6046	6060	6074	6088	6101	6115	6129	6143	2	5	7	9	12
38	.6157	6170	6184	6198	6211	6225	6239	6252	6266	6280	2	5	7	9	11
39	.6293	6307	6320	6334	6347	6361	6374	6388	6401	6414	2	4	7	9	11
40	.6428	6441	6455	6468	6481	6494	6508	6521	6534	6547	2	4	7	9	11
41	.6561	6574	6587	6600	6613	6626	6639	6652	6665	6678	2	4	7	9	11
42	.6691	6704	6717	6730	6743	6756	6769	6782	6794	6807	2	4	6	9	11
43	.6820	6833	6845	6858	6871	6884	6896	6909	6921	6934	2	4	6	8	11
44	.6947	6959	6972	6984	6997	7009	7022	7034	7046	7059	2	4	6	8	10
45	.7071	7083	7096	7108	7120	7133	7145	7157	7169	7181	2	4	6	8	10
46	.7193	7206	7218	7230	7242	7254	7266	7278	7290	7302	2	4	6	8	10
47	.7314	7325	7337	7349	7361	7373	7385	7396	7408	7420	2	4	6	8	10
48	.7431	7443	7455	7466	7478	7490	7501	7513	7524	7536	2	4	6	8	10
49	.7547	7559	7570	7581	7593	7604	7615	7627	7638	7649	2	4	6	8	9
50	.7660	7672	7683	7694	7705	7716	7727	7738	7749	7760	2	4	6	7	9
51	.7771	7782	7793	7804	7815	7826	7837	7848	7859	7869	2	4	5	7	9
52	.7880	7891	7902	7912	7923	7934	7944	7955	7965	7976	2	4	5	7	9
53	.7986	7997	8007	8018	8028	8039	8049	8059	8070	8080	2	3	5	7	9

NATURAL SINES (Continued)

Degrees	0′	6′	12′	18′	24′	30′	36′	42′	48′	54′	Mean Differences 1′	2′	3′	4′	5′
54	.8090	8100	8111	8121	8131	8141	8151	8161	8171	8181	2	3	5	7	8
55	.8192	8202	8211	8221	8231	8241	8251	8261	8271	8281	2	3	5	7	8
56	.8290	8300	8310	8320	8329	8339	8348	8358	8368	8377	2	3	5	6	8
57	.8387	8396	8406	8415	8425	8434	8443	8453	8462	8471	2	3	5	6	8
58	.8480	8490	8499	8508	8517	8526	8536	8545	8554	8563	2	3	5	6	8
59	.8572	8581	8590	8599	8607	8616	8625	8634	8643	8652	1	3	4	6	7
60	.8660	8669	8678	8686	8695	8704	8712	8721	8729	8738	1	3	4	6	7
61	.8746	8755	8763	8771	8780	8788	8796	8805	8813	8821	1	3	4	6	7
62	.8829	8838	8846	8854	8862	8870	8878	8886	8894	8902	1	3	4	5	7
63	.8910	8918	8926	8934	8942	8949	8957	8965	8973	8980	1	3	4	5	6
64	.8988	8996	9003	9011	9018	9026	9033	9041	9048	9056	1	3	4	5	6
65	.9063	9070	9078	9085	9092	9100	9107	9114	9121	9128	1	2	4	5	6
66	.9135	9143	9150	9157	9164	9171	9178	9184	9191	9198	1	2	3	5	6
67	.9205	9212	9219	9225	9232	9239	9245	9252	9259	9265	1	2	3	4	6
68	.9272	9278	9285	9291	9298	9304	9311	9317	9323	9330	1	2	3	4	5
69	.9336	9342	9348	9354	9361	9367	9373	9379	9385	9391	1	2	3	4	5
70	.9397	9403	9409	9415	9421	9426	9432	9438	9444	9449	1	2	3	4	5
71	.9455	9461	9466	9472	9478	9483	9489	9494	9500	9505	1	2	3	4	5

NATURAL SINES (Continued)

Degrees	0′	6′	12′	18′	24′	30′	36′	42′	48′	54′	Mean Differences 1′	2′	3′	4′	5′
72	.9511	9516	9521	9527	9532	9537	9542	9548	9553	9558	1	2	3	3	4
73	.9563	9568	9573	9578	9583	9588	9593	9598	9603	9608	1	2	2	3	4
74	.9613	9617	9622	9627	9632	9636	9641	9646	9650	9655	1	2	2	3	4
75	.9659	9664	9668	9673	9677	9681	9686	9690	9694	9699	1	1	2	3	4
76	.9703	9707	9711	9715	9720	9724	9728	9732	9736	9740	1	1	2	3	3
77	.9744	9748	9751	9755	9759	9763	9767	9770	9774	9778	1	1	2	3	3
78	.9781	9785	9789	9792	9796	9799	9803	9806	9810	9813	1	1	2	2	3
79	.9816	9820	9823	9826	9829	9833	9836	9839	9842	9845	1	1	2	2	3
80	.9848	9851	9854	9857	9860	9863	9866	9869	9871	9874	0	1	1	2	2
81	.9877	9880	9882	9885	9888	9890	9893	9895	9898	9900	0	1	1	2	2
82	.9903	9905	9907	9910	9912	9914	9917	9919	9921	9923	0	1	1	2	2
83	.9925	9928	9930	9932	9934	9936	9938	9940	9942	9943	0	1	1	1	2
84	.9945	9947	9949	9951	9952	9954	9956	9957	9959	9960	0	1	1	1	2
85	.9962	9963	9965	9966	9968	9969	9971	9972	9973	9974	0	0	1	1	1
86	.9976	9977	9978	9979	9980	9981	9982	9983	9984	9985	0	0	1	1	1
87	.9986	9987	9988	9989	9990	9990	9991	9992	9993	9993	0	0	0	1	1
88	.9994	9995	9995	9996	9996	9997	9997	9997	9998	9998	0	0	0	0	0
89	.9998	9999	9999	9999	9999	1.000	1.000	1.000	1.000	1.000	0	0	0	0	0

LOGARITHMIC SINES

Degrees	0′	6′	12′	18′	24′	30′	36′	42′	48′	54′	Mean Differences 1′	2′	3′	4′	5′
0	$-\infty$	7.2419	5429	7190	8439	9408	0200	0870	1450	1961					
1	8.2419	2832	3210	3558	3880	4179	4459	4723	4971	5206					
2	8.5428	5640	5842	6035	6220	6397	6567	6731	6889	7041					
3	8.7188	7330	7468	7602	7731	7857	7979	8098	8213	8326					
4	8.8436	8543	8647	8749	8849	8946	9042	9135	9226	9315	16	32	48	64	80
5	8.9403	9489	9573	9655	9736	9816	9894	9970	0046	0120	13	26	39	52	65
6	9.0192	0264	0334	0403	0472	0539	0605	0670	0734	0797	11	22	33	44	55
7	9.0859	0920	0981	1040	1099	1157	1214	1271	1326	1381	10	19	29	38	48
8	9.1436	1489	1542	1594	1646	1697	1747	1797	1847	1895	8	17	25	34	42
9	9.1943	1991	2038	2085	2131	2176	2221	2266	2310	2353	8	15	23	30	38
10	9.2397	2439	2482	2524	2565	2606	2647	2687	2727	2767	7	14	20	27	34
11	9.2806	2845	2883	2921	2959	2997	3034	3070	3107	3143	6	12	19	25	31
12	9.3179	3214	3250	3284	3319	3353	3387	3421	3455	3488	6	11	17	23	28
13	9.3521	3554	3586	3618	3650	3682	3713	3745	3775	3806	5	11	16	21	26
14	9.3837	3867	3897	3927	3957	3986	4015	4044	4073	4102	5	10	15	20	24
15	9.4130	4158	4186	4214	4242	4269	4296	4323	4350	4377	5	9	14	18	23
16	9.4403	4430	4456	4482	4508	4533	4559	4584	4609	4634	4	9	13	17	21
17	9.4659	4684	4709	4733	4757	4781	4805	4829	4853	4876	4	8	12	16	20

LOGARITHMIC SINES (Continued)

Degrees	0′	6′	12′	18′	24′	30′	36′	42′	48′	54′	Mean Differences 1′	2′	3′	4′	5′
18	9.4900	4923	4946	4969	4992	5015	5037	5060	5082	5104	4	8	11	15	19
19	9.5126	5148	5170	5192	5213	5235	5256	5278	5299	5320	4	7	11	14	18
20	9.5341	5361	5382	5402	5423	5443	5463	5484	5504	5523	3	7	10	14	17
21	9.5543	5563	5583	5602	5621	5641	5660	5679	5698	5717	3	6	10	13	16
22	9.5736	5754	5773	5792	5810	5828	5847	5865	5883	5901	3	6	9	12	15
23	9.5919	5937	5954	5972	5990	6007	6024	6042	6059	6076	3	6	9	12	15
24	9.6093	6110	6127	6144	6161	6177	6194	6210	6227	6243	3	6	8	11	14
25	9.6259	6276	6292	6308	6324	6340	6356	6371	6387	6403	3	5	8	11	13
26	9.6418	6434	6449	6465	6480	6495	6510	6526	6541	6556	3	5	8	10	13
27	9.6570	6585	6600	6615	6629	6644	6659	6673	6687	6702	2	5	7	10	12
28	9.6716	6730	6744	6759	6773	6787	6801	6814	6828	6842	2	5	7	9	12
29	9.6856	6869	6883	6896	6910	6923	6937	6950	6963	6977	2	4	7	9	11
30	9.6990	7003	7016	7029	7042	7055	7068	7080	7093	7106	2	4	6	9	11
31	9.7118	7131	7144	7156	7168	7181	7193	7205	7218	7230	2	4	6	8	10
32	9.7242	7254	7266	7278	7290	7302	7314	7326	7338	7349	2	4	6	8	10
33	9.7361	7373	7384	7396	7407	7419	7430	7442	7453	7464	2	4	6	8	10
34	9.7476	7487	7498	7509	7520	7531	7542	7553	7564	7575	2	4	6	7	9
35	9.7586	7597	7607	7618	7629	7640	7650	7661	7671	7682	2	4	5	7	9

LOGARITHMIC SINES (Continued)

Degrees	0′	6′	12′	18′	24′	30′	36′	42′	48′	54′	Mean Differences 1′	2′	3′	4′	5′
36	9.7692	7703	7713	7723	7734	7744	7754	7764	7774	7785	2	3	5	7	9
37	9.7795	7805	7815	7825	7835	7844	7854	7864	7874	7884	2	3	5	7	8
38	9.7893	7903	7913	7922	7932	7941	7951	7960	7970	7979	2	3	5	6	8
39	9.7989	7998	8007	8017	8026	8035	8044	8053	8063	8072	2	3	5	6	8
40	9.8081	8090	8099	8108	8117	8125	8134	8143	8152	8161	1	3	4	6	7
41	9.8169	8178	8187	8195	8204	8213	8221	8230	8238	8247	1	3	4	6	7
42	9.8255	8264	8272	8280	8289	8297	8305	8313	8322	8330	1	3	4	6	7
43	9.8338	8346	8354	8362	8370	8378	8386	8394	8402	8410	1	3	4	5	7
44	9.8418	8426	8433	8441	8449	8457	8464	8472	8480	8487	1	3	4	5	6
45	9.8495	8502	8510	8517	8525	8532	8540	8547	8555	8562	1	2	4	5	6
46	9.8569	8577	8584	8591	8598	8606	8613	8620	8627	8634	1	2	4	5	6
47	9.8641	8648	8655	8662	8669	8676	8683	8690	8697	8704	1	2	3	5	6
48	9.8711	8718	8724	8731	8738	8745	8751	8758	8765	8771	1	2	3	4	6
49	9.8778	8784	8791	8797	8804	8810	8817	8823	8830	8836	1	2	3	4	5
50	9.8843	8849	8855	8862	8868	8874	8880	8887	8893	8899	1	2	3	4	5
51	9.8905	8911	8917	8923	8929	8935	8941	8947	8953	8959	1	2	3	4	5
52	9.8965	8971	8977	8983	8989	8995	9000	9006	9012	9018	1	2	3	4	5
53	9.9023	9029	9035	9041	9046	9052	9057	9063	9069	9074	1	2	3	4	5

LOGARITHMIC SINES (Continued)

Degrees	0′	6′	12′	18′	24′	30′	36′	42′	48′	54′	Mean Differences 1′	2′	3′	4′	5′
54	9.9080	9085	9091	9096	9101	9107	9112	9118	9123	9128	1	2	3	4	5
55	9.9134	9139	9144	9149	9155	9160	9165	9170	9175	9181	1	2	3	3	4
56	9.9186	9191	9196	9201	9206	9211	9216	9221	9226	9231	1	2	3	3	4
57	9.9236	9241	9246	9251	9255	9260	9265	9270	9275	9279	1	2	2	3	4
58	9.9284	9289	9294	9298	9303	9308	9312	9317	9322	9326	1	2	2	3	4
59	9.9331	9335	9340	9344	9349	9351	9358	9362	9367	9371	1	1	2	3	4
60	9.9375	9380	9384	9388	9393	9397	9401	9406	9410	9414	1	1	2	3	4
61	9.9418	9422	9427	9431	9435	9439	9443	9447	9451	9455	1	1	2	3	3
62	9.9459	9463	9467	9471	9475	9479	9483	9487	9491	9495	1	1	2	3	3
63	9.9499	9503	9507	9510	9514	9518	9522	9525	9529	9533	1	1	2	3	3
64	9.9537	9540	9544	9548	9551	9555	9558	9562	9566	9569	1	1	2	2	3
65	9.9573	9576	9580	9583	9587	9590	9594	9597	9601	9604	1	1	2	2	3
66	9.9607	9611	9614	9617	9621	9624	9627	9631	9634	9637	1	1	2	2	3
67	9.9640	9643	9647	9650	9653	9656	9659	9662	9666	9669	1	1	2	2	3
68	9.9672	9675	9678	9681	9684	9687	9690	9693	9696	9699	0	1	1	2	2
69	9.9702	9704	9707	9710	9713	9716	9719	9722	9724	9727	0	1	1	2	2
70	9.9730	9733	9735	9738	9741	9743	9746	9749	9751	9754	0	1	1	2	2
71	9.9757	9759	9762	9764	9767	9770	9772	9775	9777	9780	0	1	1	2	2

LOGARITHMIC SINES (Continued)

Degrees	0′	6′	12′	18′	24′	30′	36′	42′	48′	54′	Mean Differences 1′	2′	3′	4′	5′
72	9.9782	9785	9787	9789	9792	9794	9797	9799	9801	9804	0	1	1	2	2
73	9.9806	9808	9811	9813	9815	9817	9820	9822	9824	9826	0	1	1	2	2
74	9.9828	9831	9833	9835	9837	9839	9841	9843	9845	9847	0	1	1	1	2
75	9.9849	9851	9853	9855	9857	9859	9861	9863	9865	9867	0	1	1	1	2
76	9.9869	9871	9873	9875	9876	9878	9880	9882	9884	9885	0	1	1	1	2
77	9.9887	9889	9891	9892	9894	9896	9897	9899	9901	9902	0	1	1	1	1
78	9.9904	9906	9907	9909	9910	9912	9913	9915	9916	9918	0	1	1	1	1
79	9.9919	9921	9922	9924	9925	9927	9928	9929	9931	9932	0	0	1	1	1
80	9.9934	9935	9936	9937	9939	9940	9941	9943	9944	9945	0	0	1	1	1
81	9.9946	9947	9949	9950	9951	9952	9953	9954	9955	9956	0	0	1	1	1
82	9.9958	9959	9960	9961	9962	9963	9964	9965	9966	9967	0	0	1	1	1
83	9.9968	9968	9969	9970	9971	9972	9973	9974	9975	9975	0	0	0	0	1
84	9.9976	9977	9978	9978	9979	9980	9981	9981	9982	9983	0	0	0	0	1
85	9.9983	9984	9985	9985	9986	9987	9987	9988	9988	9989	0	0	0	0	0
86	9.9989	9990	9990	9991	9991	9992	9992	9993	9993	9994	0	0	0	0	0
87	9.9994	9994	9995	9995	9996	9996	9996	9996	9997	9997	0	0	0	0	0
88	9.9997	9998	9998	9998	9998	9999	9999	9999	9999	9999	0	0	0	0	0
89	9.9999	9999	10.00	10.00	10.00	10.00	10.00	10.00	10.00	10.00	0	0	0	0	0

NATURAL COSINES

Degrees	0′	6′	12′	18′	24′	30′	36′	42′	48′	54′	Mean Differences 1′	2′	3′	4′	5′
0	1.000	1.000	1.000	1.000	1.000	1.000	.9999	9999	9999	9999	0	0	0	0	0
1	.9998	9998	9998	9997	9997	9997	9996	9996	9995	9995	0	0	0	0	0
2	.9994	9993	9993	9992	9991	9990	9990	9989	9988	9987	0	0	0	1	1
3	.9986	9985	9984	9983	9982	9981	9980	9979	9978	9977	0	0	1	1	1
4	.9976	9974	9973	9972	9971	9969	9968	9966	9965	9963	0	0	1	1	1
5	.9962	9960	9959	9957	9956	9954	9952	9951	9949	9947	0	1	1	1	2
6	.9945	9943	9942	9940	9938	9936	9934	9932	9930	9928	0	1	1	1	2
7	.9925	9923	9921	9919	9917	9914	9912	9910	9907	9905	0	1	1	2	2
8	.9903	9900	9898	9895	9893	9890	9888	9885	9882	9880	0	1	1	2	2
9	.9877	9874	9871	9869	9866	9863	9860	9857	9854	9851	0	1	1	2	2
10	.9848	9845	9842	9839	9836	9833	9829	9826	9823	9820	1	1	2	2	3
11	.9816	9813	9810	9806	9803	9799	9796	9792	9789	9785	1	1	2	2	3
12	.9781	9778	9774	9770	9767	9763	9759	9755	9751	9748	1	1	2	3	3
13	.9744	9740	9736	9732	9728	9724	9720	9715	9711	9707	1	1	2	3	3
14	.9703	9699	9694	9690	9686	9681	9677	9673	9668	9664	1	1	2	3	4
15	.9659	9655	9650	9646	9641	9636	9632	9627	9622	9617	1	2	2	3	4
16	.9613	9608	9603	9598	9593	9588	9583	9578	9573	9568	1	2	2	3	4
17	.9563	9558	9553	9548	9542	9537	9532	9527	9521	9516	1	2	3	3	4

N.B.—Subtract Mean Differences.

NATURAL COSINES (Continued)

Degrees	0′	6′	12′	18′	24′	30′	36′	42′	48′	54′	Mean Differences 1′	2′	3′	4′	5′
18	.9511	9505	9500	9494	9489	9483	9478	9472	9466	9461	1	2	3	4	5
19	.9455	9449	9444	9438	9432	9426	9421	9415	9409	9403	1	2	3	4	5
20	.9397	9391	9385	9379	9373	9367	9361	9354	9348	9342	1	2	3	4	5
21	.9336	9330	9323	9317	9311	9304	9298	9291	9285	9278	1	2	3	4	5
22	.9272	9265	9259	9252	9245	9239	9232	9225	9219	9212	1	2	3	4	6
23	.9205	9198	9191	9184	9178	9171	9164	9157	9150	9143	1	2	3	5	6
24	.9135	9128	9121	9114	9107	9100	9092	9085	9078	9070	1	2	4	5	6
25	.9063	9056	9048	9041	9033	9026	9018	9011	9003	8996	1	3	4	5	6
26	.8988	8980	8973	8965	8957	8949	8942	8934	8926	8918	1	3	4	5	6
27	.8910	8902	8894	8886	8878	8870	8862	8854	8846	8838	1	3	4	5	7
28	.8829	8821	8813	8805	8796	8788	8780	8771	8763	8755	1	3	4	6	7
29	.8746	8738	8729	8721	8712	8704	8695	8686	8678	8669	1	3	4	6	7
30	.8660	8652	8643	8634	8625	8616	8607	8599	8590	8581	1	3	4	6	7
31	.8572	8563	8554	8545	8536	8526	8517	8508	8499	8490	2	3	5	6	8
32	.8480	8471	8462	8453	8443	8434	8425	8415	8406	8396	2	3	5	6	8
33	.8387	8377	8368	8358	8348	8339	8329	8320	8310	8300	2	3	5	6	8
34	.8290	8281	8271	8261	8251	8241	8231	8221	8211	8202	2	3	5	7	8
35	.8192	8181	8171	8161	8151	8141	8131	8121	8111	8100	2	3	5	7	8

N.B.—Subtract Mean Differences.

NATURAL COSINES (Continued)

Deg.	0′	6′	12′	18′	24′	30′	36′	42′	48′	54′	Mean Differences 1′	2′	3′	4′	5′
36	.8090	8080	8070	8059	8049	8039	8028	8018	8007	7997	2	3	5	7	9
37	.7986	7976	7965	7955	7944	7934	7923	7912	7902	7891	2	4	5	7	9
38	.7880	7869	7859	7848	7837	7826	7815	7804	7793	7782	2	4	5	7	9
39	.7771	7760	7749	7738	7728	7716	7705	7694	7683	7672	2	4	6	7	9
40	.7660	7649	7638	7627	7615	7604	7593	7581	7570	7559	2	4	6	8	9
41	.7547	7536	7524	7513	7501	7490	7478	7466	7455	7443	2	4	6	8	10
42	.7431	7420	7408	7396	7385	7373	7361	7349	7337	7325	2	4	6	8	10
43	.7314	7302	7290	7278	7266	7254	7242	7230	7218	7206	2	4	6	8	10
44	.7193	7181	7169	7157	7145	7133	7120	7108	7096	7083	2	4	6	8	10
45	.7071	7059	7046	7034	7022	7009	6997	6984	6972	6959	2	4	6	8	10
46	.6947	6934	6921	6909	6896	6884	6871	6858	6845	6833	2	4	6	8	11
47	.6820	6807	6794	6782	6769	6756	6743	6730	6717	6704	2	4	6	9	11
48	.6691	6678	6665	6652	6639	6626	6613	6600	6587	6574	2	4	7	9	11
49	.6561	6547	6534	6521	6508	6494	6481	6468	6455	6441	2	4	7	9	11
50	.6428	6414	6401	6388	6374	6361	6347	6334	6320	6307	2	4	7	9	11
51	.6293	6280	6266	6252	6239	6225	6211	6198	6184	6170	2	5	7	9	11
52	.6157	6143	6129	6115	6101	6088	6074	6060	6046	6032	2	5	7	9	12
53	.6018	6004	5990	5976	5962	5948	5934	5920	5906	5892	2	5	7	9	12

N.B.—Subtract Mean Differences.

NATURAL COSINES (Continued)

Deg.	0′	6′	12′	18′	24′	30′	36′	42′	48′	54′	Mean Differences 1′	2′	3′	4′	5′
54	.5878	5864	5850	5835	5821	5807	5793	5779	5764	5750	2	5	7	9	12
55	.5736	5721	5707	5693	5678	5661	5650	5635	5621	5606	2	5	7	10	12
56	.5592	5577	5563	5548	5534	5519	5505	5490	5476	5461	2	5	7	10	12
57	.5446	5432	5417	5402	5388	5373	5358	5344	5329	5314	2	5	7	10	12
58	.5299	5284	5270	5255	5240	5225	5210	5195	5180	5165	2	5	7	10	12
59	.5150	5135	5120	5105	5090	5075	5060	5045	5030	5015	3	5	8	10	13
60	.5000	4985	4970	4955	4939	4924	4909	4894	4879	4863	3	5	8	10	13
61	.4848	4833	4818	4802	4787	4772	4756	4741	4726	4710	3	5	8	10	13
62	.4695	4679	4664	4648	4633	4617	4602	4586	4571	4555	3	5	8	10	13
63	.4540	4524	4509	4493	4478	4462	4446	4431	4415	4399	3	5	8	10	13
64	.4384	4368	4352	4337	4321	4305	4289	4274	4258	4242	3	5	8	11	13
65	.4226	4210	4195	4179	4163	4147	4131	4115	4099	4083	3	5	8	11	13
66	.4067	4051	4035	4019	4003	3987	3971	3955	3939	3923	3	5	8	11	14
67	.3907	3891	3875	3859	3843	3827	3811	3795	3778	3762	3	5	8	11	14
68	.3746	3730	3714	3697	3681	3665	3649	3633	3616	3600	3	5	8	11	14
69	.3584	3567	3551	3535	3518	3502	3486	3469	3453	3437	3	5	8	11	14
70	.3420	3404	3387	3371	3355	3338	3322	3305	3289	3272	3	5	8	11	14
71	.3256	3239	3223	3206	3190	3173	3156	3140	3123	3107	3	6	8	11	14

N.B.—Subtract Mean Differences.

NATURAL COSINES (Continued)

Deg.	0′	6′	12′	18′	24′	30′	36′	42′	48′	54′	Mean Differences 1′	2′	3′	4′	5′
72	.3090	3074	3057	3040	3024	3007	2990	2974	2957	2940	3	6	8	11	14
73	.2924	2907	2890	2874	2857	2840	2823	2807	2790	2773	3	6	8	11	14
74	.2756	2740	2723	2706	2689	2672	2656	2639	2622	2605	3	6	8	11	14
75	.2588	2571	2554	2538	2521	2504	2487	2470	2453	2436	3	6	8	11	14
76	.2419	2402	2385	2368	2351	2334	2317	2300	2284	2267	3	6	8	11	14
77	.2250	2233	2215	2198	2181	2164	2147	2130	2113	2096	3	6	9	11	14
78	.2079	2062	2045	2028	2011	1994	1977	1959	1942	1925	3	6	9	11	14
79	.1908	1891	1874	1857	1840	1822	1805	1788	1771	1754	3	6	9	11	14
80	.1736	1719	1702	1685	1668	1650	1633	1616	1599	1582	3	6	9	12	14
81	.1564	1547	1530	1513	1495	1478	1461	1444	1426	1409	3	6	9	12	14
82	.1392	1374	1357	1340	1323	1305	1288	1271	1253	1236	3	6	9	12	14
83	.1219	1201	1184	1167	1149	1132	1115	1097	1080	1063	3	6	9	12	14
84	.1045	1028	1011	0993	0976	0958	0941	0924	0906	0889	3	6	9	12	14
85	.0872	0854	0837	0819	0802	0785	0767	0750	0732	0715	3	6	9	12	14
86	.0698	0680	0663	0645	0628	0610	0593	0576	0558	0541	3	6	9	12	15
87	.0523	0506	0488	0471	0454	0436	0419	0401	0384	0366	3	6	9	12	15
88	.0349	0332	0314	0297	0279	0262	0244	0227	0209	0192	3	6	9	12	15
89	.0175	0157	0140	0122	0105	0087	0070	0052	0035	0017	3	6	9	12	15

N.B.—Subtract Mean Differences

LOGARITHMIC COSINES

Deg.	0′	6′	12′	18′	24′	30′	36′	42′	48′	54′	Mean Differences 1′	2′	3′	4′	5′
0	10.0000	0000	0000	0000	0000	0000	0000	0000	0000	9.9999	0	0	0	0	0
1	9.9999	9999	9999	9999	9999	9999	9998	9998	9998	9998	0	0	0	0	0
2	9.9997	9997	9997	9996	9996	9996	9996	9995	9995	9994	0	0	0	0	0
3	9.9994	9994	9993	9993	9992	9992	9991	9991	9990	9990	0	0	0	0	0
4	9.9989	9989	9988	9988	9987	9987	9986	9985	9985	9984	0	0	0	0	0
5	9.9983	9983	9982	9981	9981	9980	9979	9978	9978	9977	0	0	0	0	1
6	9.9976	9975	9975	9974	9973	9972	9971	9970	9969	9968	0	0	0	1	1
7	9.9968	9967	9966	9965	9964	9963	9962	9961	9960	9959	0	0	1	1	1
8	9.9958	9956	9955	9954	9953	9952	9951	9950	9949	9947	0	0	1	1	1
9	9.9946	9945	9944	9943	9941	9940	9939	9937	9936	9935	0	0	1	1	1
10	9.9934	9932	9931	9929	9928	9927	9925	9924	9922	9921	0	0	1	1	1
11	9.9919	9918	9916	9915	9913	9912	9910	9909	9907	9906	0	1	1	1	1
12	9.9904	9902	9901	9899	9897	9896	9894	9892	9891	9889	0	1	1	1	1
13	9.9887	9885	9884	9882	9880	9878	9876	9875	9873	9871	0	1	1	1	2
14	9.9869	9867	9865	9863	9861	9859	9857	9855	9853	9851	0	1	1	1	2

N.B.—Subtract Mean Differences.

LOGARITHMIC COSINES (Continued)

Deg.	0′	6′	12′	18′	24′	30′	36′	42′	48′	54′	Mean Differences 1′	2′	3′	4′	5′
15	9.9849	9847	9845	9843	9841	9839	9837	9835	9833	9831	0	1	1	1	2
16	9.9828	9826	9824	9822	9820	9817	9815	9813	9811	9808	0	1	1	2	2
17	9.9806	9804	9801	9799	9797	9794	9792	9798	9787	9785	0	1	1	2	2
18	9.9782	9780	9777	9775	9772	9770	9767	9764	9762	9759	0	1	1	2	2
19	9.9757	9754	9751	9749	9746	9743	9741	9738	9735	9733	0	1	1	2	2
20	9.9730	9727	9724	9722	9719	9716	9713	9710	9707	9704	0	1	1	2	2
21	9.9702	9699	9696	9693	9690	9687	9684	9681	9678	9675	0	1	1	2	2
22	9.9672	9669	9666	9662	9659	9656	9653	9650	9647	9643	1	1	2	2	3
23	9.9640	9637	9634	9631	9627	9624	9621	9617	9614	9611	1	1	2	2	3
24	9.9607	9604	9601	9597	9594	9590	9587	9583	9580	9576	1	1	2	2	3
25	9.9573	9569	9566	9562	9558	9555	9551	9548	9544	9540	1	1	2	2	3
26	9.9537	9533	9529	9525	9522	9518	9514	9510	9507	9503	1	1	2	3	3
27	9.9499	9495	9491	9487	9483	9479	9475	9471	9467	9463	1	1	2	3	3
28	9.9459	9455	9451	9447	9443	9439	9435	9431	9427	9422	1	1	2	3	3

N.B.—Subtract Mean Differences.

LOGARITHMIC COSINES (Continued)

Deg.	0′	6′	12′	18′	24′	30′	36′	42′	48′	54′	Mean Differences 1′	2′	3′	4′	5′
29	9.9418	9414	9410	9406	9401	9397	9393	9388	9384	9380	1	1	2	3	4
30	9.9375	9371	9367	9362	9358	9353	9349	9344	9340	9335	1	1	2	3	4
31	9.9331	9326	9322	9317	9312	9308	9303	9298	9294	9289	1	2	2	3	4
32	9.9284	9279	9275	9270	9265	9260	9255	9251	9246	9241	1	2	2	3	4
33	9.9236	9231	9226	9221	9216	9211	9206	9201	9196	9191	1	2	3	3	4
34	9.9186	9181	9175	9170	9165	9160	9155	9149	9144	9139	1	2	3	3	4
35	9.9134	9128	9123	9118	9112	9107	9101	9096	9091	9085	1	2	3	4	5
36	9.9080	9074	9069	9063	9057	9052	9046	9041	9035	9029	1	2	3	4	5
37	9.9023	9018	9012	9006	9000	8995	8989	8983	8977	8971	1	2	3	4	5
38	9.8965	8959	8953	8947	8941	8935	8929	8923	8917	8911	1	2	3	4	5
39	9.8905	8899	8893	8887	8880	8874	8868	8862	8855	8849	1	2	3	4	5
40	9.8843	8836	8830	8823	8817	8810	8804	8797	8791	8784	1	2	3	4	5
41	9.8778	8771	8765	8758	8751	8745	8738	8731	8724	8718	1	2	3	5	6
42	9.8711	8704	8697	8690	8683	8676	8669	8662	8655	8648	1	2	3	5	6
43	9.8641	8634	8627	8620	8613	8606	8598	8591	8584	8577	1	2	4	5	6
44	9.8569	8562	8555	8547	8540	8532	8525	8517	8510	8502	1	2	4	5	6

N.B.—Subtract Mean Differences.

LOGARITHMIC COSINES (Continued)

Deg.	0′	6′	12′	18′	24′	30′	36′	42′	48′	54′	Mean Differences 1′	2′	3′	4′	5′
45	9.8495	8487	8480	8472	8464	8457	8449	8441	8433	8426	1	3	4	5	6
46	9.8418	8410	8402	8394	8386	8378	8370	8362	8354	8346	1	3	4	5	7
47	9.8338	8330	8322	8313	8305	8297	8289	8280	8272	8264	1	3	4	6	7
48	9.8255	8247	8238	8230	8221	8213	8204	8195	8187	8178	1	3	4	6	7
49	9.8169	8161	8152	8143	8134	8125	8117	8108	8099	8090	1	3	4	6	7
50	9.8081	8072	8063	8053	8044	8035	8026	8017	8007	7998	2	3	5	6	8
51	9.7989	7979	7970	7960	7951	7941	7932	7922	7913	7903	2	3	5	6	8
52	9.7893	7884	7874	7864	7854	7844	7835	7825	7815	7805	2	3	5	7	8
53	9.7795	7785	7774	7764	7754	7744	7734	7723	7713	7703	2	3	5	7	9
54	9.7692	7682	7671	7661	7650	7640	7629	7618	7607	7597	2	4	5	7	9
55	9.7586	7575	7564	7553	7542	7531	7520	7509	7498	7487	2	4	6	7	9
56	9.7476	7464	7453	7442	7430	7419	7407	7396	7384	7373	2	4	6	8	10
57	9.7361	7349	7338	7326	7314	7302	7290	7278	7266	7254	2	4	6	8	10
58	9.7242	7230	7218	7205	7193	7181	7168	7156	7144	7131	2	4	6	8	10
59	9.7118	7106	7093	7080	7068	7055	7042	7029	7016	7003	2	4	6	9	11

N.B.—Subtract Mean Differences.

LOGARITHMIC COSINES (Continued)

Deg.	0′	6′	12′	18′	24′	30′	36′	42′	48′	54′	Mean Differences 1′	2′	3′	4′	5′
60	9.6990	6977	6963	6950	6937	6923	6910	6896	6883	6869	2	4	7	9	11
61	9.6856	6842	6828	6814	6801	6787	6773	6759	6744	6730	2	5	7	9	12
62	9.6716	6702	6687	6673	6659	6644	6629	6615	6600	6585	2	5	7	10	12
63	9.6570	6556	6541	6526	6510	6495	6480	6465	6449	6434	3	5	8	10	13
64	9.6418	6403	6387	6371	6356	6340	6324	6308	6292	6276	3	5	8	11	13
65	9.6259	6243	6227	6210	6194	6177	6161	6144	6127	6110	3	6	8	11	14
66	9.6093	6076	6059	6042	6024	6007	5990	5972	5954	5937	3	6	9	12	15
67	9.5919	5901	5883	5865	5847	5828	5810	5792	5773	5754	3	6	9	12	15
68	9.5736	5717	5698	5679	5660	5641	5621	5602	5583	5563	3	6	10	13	16
69	9.5543	5523	5504	5484	5463	5443	5423	5402	5382	5361	3	7	10	14	17
70	9.5341	5320	5299	5278	5256	5235	5213	5192	5170	5148	4	7	11	14	18
71	9.5126	5104	5082	5060	5037	5015	4992	4969	4946	4923	4	8	11	15	19
72	9.4900	4876	4853	4829	4805	4781	4757	4733	4709	4684	4	8	12	16	20
73	9.4659	4634	4609	4584	4559	4533	4508	4482	4456	4430	4	9	13	17	21
74	9.4403	4377	4350	4323	4296	4269	4242	4214	4186	4158	5	9	14	18	23

N.B.—Subtract Mean Differences.

LOGARITHMIC COSINES (Continued)

Deg.	0′	6′	12′	18	24′	30′	36′	42′	48′	54′	Mean Differences 1′	2′	3′	4′	5′
75	9.4130	4102	4073	4044	4015	3986	3957	3927	3897	3867	5	10	15	20	24
76	9.3837	3806	3775	3745	3713	3682	3650	3618	3586	3554	5	11	16	21	26
77	9.3521	3488	3455	3421	3387	3353	3319	3284	3250	3214	6	11	17	23	28
78	9.3179	3143	3107	3070	3034	2997	2959	2921	2883	2845	6	12	19	25	31
79	9.2806	2767	2727	2687	2647	2606	2565	2524	2482	2439	7	14	20	27	34
80	9.2397	2353	2310	2266	2221	2176	2131	2085	2038	1991	8	15	23	30	38
81	9.1943	1895	1847	1797	1747	1697	1646	1594	1542	1489	8	17	25	34	42
82	9.1436	1381	1326	1271	1214	1157	1099	1040	0981	0920	10	19	29	38	48
83	9.0859	0797	0734	0670	0605	0539	0472	0403	0334	0264	11	22	33	44	55
84	9.0192	0120	0046	$\overline{9970}$	$\overline{9894}$	$\overline{9816}$	$\overline{9736}$	$\overline{9655}$	$\overline{9573}$	$\overline{9489}$	13	26	39	52	65
85	8.9403	9315	9226	9135	9042	8946	8849	8749	8647	8543	16	32	48	64	80
86	8.8436	8326	8213	8098	7979	7857	7731	7602	7468	7330	*Mean* differences no longer sufficiently accurate.				
87	8.7188	7041	6889	6731	6567	6397	6220	6035	5842	5640					
88	8.5428	5206	4971	4723	4459	4179	3880	3558	3210	2832					
89	8.2419	1961	1450	0870	0200	$\overline{9408}$	$\overline{8439}$	$\overline{7190}$	$\overline{5429}$	$\overline{2419}$					

N.B.—Subtract Mean Differences.

NATURAL TANGENTS

Deg.	0′	6′	12′	18′	24′	30′	36′	42′	48′	54′	Mean Differences 1′	2′	3′	4′	5′
0	.0000	0017	0035	0052	0070	0087	0105	0122	0140	0157	3	6	9	12	15
1	.0175	0192	0209	0227	0244	0262	0279	0297	0314	0332	3	6	9	12	15
2	.0349	0367	0384	0402	0419	0437	0454	0472	0489	0507	3	6	9	12	15
3	.0524	0542	0559	0577	0594	0612	0629	0647	0664	0682	3	6	9	12	15
4	.0699	0717	0734	0752	0769	0787	0805	0822	0840	0857	3	6	9	12	15
5	.0875	0892	0910	0928	0945	0963	0981	0998	1016	1033	3	6	9	12	15
6	.1051	1069	1086	1104	1122	1139	1157	1175	1192	1210	3	6	9	12	15
7	.1228	1246	1263	1281	1299	1317	1334	1352	1370	1388	3	6	9	12	15
8	.1405	1423	1441	1459	1477	1495	1512	1530	1548	1566	3	6	9	12	15
9	.1584	1602	1620	1638	1655	1673	1691	1709	1727	1745	3	6	9	12	15
10	.1763	1781	1799	1817	1835	1853	1871	1890	1908	1926	3	6	9	12	15
11	.1944	1962	1980	1998	2016	2035	2053	2071	2089	2107	3	6	9	12	15
12	.2126	2144	2162	2180	2199	2217	2235	2254	2272	2290	3	6	9	12	15
13	.2309	2327	2345	2364	2382	2401	2419	2438	2456	2475	3	6	9	12	15
14	.2493	2512	2530	2549	2568	2586	2605	2623	2642	2661	3	6	9	12	16
15	.2679	2698	2717	2736	2754	2773	2792	2811	2830	2849	3	6	9	13	16
16	.2867	2886	2905	2924	2943	2962	2981	3000	3019	3038	3	6	9	13	16
17	.3057	3076	3096	3115	3134	3153	3172	3191	3211	3230	3	6	10	13	16

NATURAL TANGENTS (Continued)

Deg.	0′	6′	12′	18′	24′	30′	36′	42′	48′	54′	Mean Differences 1′	2′	3′	4′	5′
18	.3249	3269	3288	3307	3327	3346	3365	3385	3404	3424	3	6	10	13	16
19	.3443	3463	3482	3502	3522	3541	3561	3581	3600	3620	3	7	10	13	16
20	.3640	3659	3679	3699	3719	3739	3759	3779	3799	3819	3	7	10	13	17
21	.3839	3859	3879	3899	3919	3939	3959	3979	4000	4020	3	7	10	13	17
22	.4040	4061	4081	4101	4122	4142	4163	4183	4204	4224	3	7	10	14	17
23	.4245	4265	4286	4307	4327	4348	4369	4390	4411	4431	3	7	10	14	17
24	.4452	4473	4494	4515	4536	4557	4578	4599	4621	4642	4	7	11	14	18
25	.4663	4684	4706	4727	4748	4770	4791	4813	4834	4856	4	7	11	14	18
26	.4877	4899	4921	4942	4964	4986	5008	5029	5051	5073	4	7	11	15	18
27	.5095	5117	5139	5161	5184	5206	5228	5250	5272	5295	4	7	11	15	18
28	.5317	5340	5362	5384	5407	5430	5452	5475	5498	5520	4	8	11	15	19
29	.5543	5566	5589	5612	5635	5658	5681	5704	5727	5750	4	8	12	15	19
30	.5774	5797	5820	5844	5867	5890	5914	5938	5961	5985	4	8	12	16	20
31	.6009	6032	6056	6080	6104	6128	6152	6176	6200	6224	4	8	12	16	20
32	.6249	6273	6297	6322	6346	6371	6395	6420	6445	6469	4	8	12	16	20
33	.6494	6519	6544	6569	6594	6619	6644	6669	6694	6720	4	8	13	17	21
34	.6745	6771	6796	6822	6847	6873	6899	6924	6950	6976	4	9	13	17	21
35	.7002	7028	7054	7080	7107	7133	7159	7186	7212	7239	4	9	13	18	22

NATURAL TANGENTS (Continued)

Deg.	0′	6′	12′	18′	24′	30′	36′	42′	48′	54′	Mean Differences 1′	2′	3′	4′	5′
36	.7265	7292	7319	7346	7373	7400	7427	7454	7481	7508	5	9	14	18	23
37	.7536	7563	7590	7618	7646	7673	7701	7729	7757	7785	5	9	14	18	23
38	.7813	7841	7869	7898	7926	7954	7983	8012	8040	8069	5	9	14	19	24
39	.8098	8127	8156	8185	8214	8243	8273	8302	8332	8361	5	10	15	20	24
40	.8391	8421	8451	8481	8511	8541	8571	8601	8632	8662	5	10	15	20	25
41	.8693	8724	8754	8785	8816	8847	8878	8910	8941	8972	5	10	16	21	26
42	.9004	9036	9067	9099	9131	9163	9195	9228	9260	9293	5	11	16	21	27
43	.9325	9358	9391	9424	9457	9490	9523	9556	9590	9623	6	11	17	22	28
44	.9657	9691	9725	9759	9793	9827	9861	9896	9930	9965	6	11	17	23	29
45	1.0000	0035	0070	0105	0141	0176	0212	0247	0283	0319	6	12	18	24	30
46	1.0355	0392	0428	0464	0501	0538	0575	0612	0649	0686	6	12	18	25	31
47	1.0724	0761	0799	0837	0875	0913	0951	0990	1028	1067	6	13	19	25	32
48	1.1106	1145	1184	1224	1263	1303	1343	1383	1423	1463	7	13	20	27	33
49	1.1504	1544	1585	1626	1667	1708	1750	1792	1833	1875	7	14	21	28	34
50	1.1918	1960	2002	2045	2088	2131	2174	2218	2261	2305	7	14	22	29	36
51	1.2349	2393	2437	2482	2527	2572	2617	2662	2708	2573	8	15	23	30	38
52	1.2799	2846	2892	2938	2985	3032	3079	3127	3175	3222	8	16	24	31	39
53	1.3270	3319	3367	3416	3465	3514	3564	3613	3663	3713	8	16	25	33	41

NATURAL TANGENTS (Continued)

Deg.	0′	6′	12′	18′	24′	30′	36′	42′	48′	54′	Mean Differences 1′	2′	3′	4′	5′
54	1.3764	3814	3865	3916	3968	4019	4071	4124	4176	4229	9	17	26	34	43
55	1.4281	4335	4388	4442	4496	4550	4605	4659	4715	4770	9	18	27	36	45
56	1.4826	4882	4938	4994	5051	5108	5166	5224	5282	5340	10	19	29	38	48
57	1.5399	5458	5517	5577	5637	5697	5757	5818	5880	5941	10	20	30	40	50
58	1.6003	6066	6128	6191	6255	6319	6383	6447	6512	6577	11	21	32	43	53
59	1.6643	6709	6775	6842	6909	6977	7045	7113	7182	7251	11	23	34	45	56
60	1.7321	7391	7461	7532	7603	7675	7747	7820	7893	7966	12	24	36	48	60
61	1.8040	8115	8190	8265	8341	8418	8495	8572	8650	8728	13	26	38	51	64
62	1.8807	8887	8967	9047	9128	9210	9292	9375	9458	9542	14	27	41	55	68
63	1.9626	9711	9797	9883	9970	0057	0145	0233	0323	0413	15	29	44	58	73
64	2.0503	0594	0686	0778	0872	0965	1060	1155	1251	1348	16	31	47	63	78
65	2.1445	1543	1642	1742	1842	1943	2045	2148	2251	2355	17	34	51	68	85
66	2.2460	2566	2673	2781	2889	2998	3109	3220	3332	3445	18	37	55	73	92
67	2.3559	3673	3789	3906	4023	4142	4262	4383	4504	4627	20	40	60	79	99
68	2.4751	4876	5002	5129	5257	5386	5517	5649	5782	5916	22	43	65	87	108
69	2.6051	6187	6325	6464	6605	6746	6889	7034	7179	7326	24	47	71	95	119
70	2.7475	7625	7776	7929	8083	8239	8397	8556	8716	8878	26	52	78	104	131
71	2.9042	9208	9375	9544	9714	9887	0061	0237	0415	0595	29	58	87	116	145

NATURAL TANGENTS (Continued)

Deg.	0′	6′	12′	18′	24′	30′	36′	42′	48′	54′	Mean Differences 1′	2′	3′	4′	5′
72	3.0777	0961	1146	1334	1524	1716	1910	2106	2305	2506	32	64	96	129	161
73	3.2709	2914	3122	3332	3544	3759	3977	4197	4420	4646	36	72	108	144	180
74	3.4874	5105	5339	5576	5816	6059	6305	6554	6806	7062	41	81	122	163	204
75	3 7321	7583	7848	8118	8391	8667	8947	9232	9520	9812	46	93	139	186	232
76	4.0108	0408	0713	1022	1335	1653	1976	2303	2635	2972					
77	4.3315	3662	4015	4374	4737	5107	5483	5864	6252	6646					
78	4.7046	7453	7867	8288	8716	9152	95[illegible]4	$\overline{0045}$	$\overline{0504}$	$\overline{0970}$					
79	5.1446	1929	2422	2924	3435	3955	4486	5026	5578	6140					
80	5.6713	7297	7894	8502	9124	9758	$\overline{0405}$	$\overline{1066}$	$\overline{1742}$	$\overline{2432}$					
81	6.3138	3859	4596	5350	6122	6912	7720	8548	9395	$\overline{0264}$	*Mean* differences no longer sufficiently accurate.				
82	7.1154	2066	3002	3962	4947	5958	6996	8062	9158	$\overline{0285}$					
83	8.1443	2636	3863	5126	6427	7769	9152	$\overline{0579}$	$\overline{2052}$	$\overline{3572}$					
84	9.5144	9.677	9.845	10.02	10.20	10.39	10.58	10.78	10.99	11.20					
85	11.430	11.66	11.91	12.16	12.43	12.71	13.00	13.30	13.62	13.95					
86	14.301	14.67	15.06	15.46	15.89	16.35	16.83	17.34	17.89	18.46					
87	19.081	19.74	20.45	21.20	22.02	22.90	23.86	24.90	26.03	27.27					
88	28.636	30.14	31.82	33.69	35.80	38.19	40.92	44.07	47.74	52.08					
89	57.290	63.66	71.62	81.85	95.49	114.6	143.2	191.0	286.5	573.0					

LOGARITHMIC TANGENTS

Deg.	0′	6′	12′	18′	24′	30′	36′	42′	48′	54′	Mean Differences 1′	2′	3′	4′	5′
0	—∞	7.2419	5429	7190	8439	9409	0200	0870	1450	1962					
1	8.2419	2833	3211	3559	3881	4181	4461	4725	4973	5208					
2	8.5431	5643	5845	6038	6223	6401	6571	6736	6894	7046					
3	8.7194	7337	7475	7609	7739	7865	7988	8107	8223	8336					
4	8.8446	8554	8659	8762	8862	8960	9056	9150	9241	9331	16	32	48	64	81
5	8.9420	9506	9591	9674	9756	9836	9915	9992	0068	0143	13	26	40	53	66
6	9.0216	0289	0360	0430	0499	0567	0633	0699	0764	0828	11	22	34	45	56
7	9.0891	0954	1015	1076	1135	1194	1252	1310	1367	1423	10	20	29	39	49
8	9.1478	1533	1587	1640	1693	1745	1797	1848	1898	1948	9	17	26	35	43
9	9.1997	2046	2094	2142	2189	2236	2282	2328	2374	2419	8	16	23	31	39
10	9.2463	2507	2551	2594	2637	2680	2722	2764	2805	2846	7	14	21	28	35
11	9.2887	2927	2967	3006	3046	3085	3123	3162	3200	3237	6	13	19	26	32
12	9.3275	3312	3349	3385	3422	3458	3493	3529	3564	3599	6	12	18	24	30
13	9.3634	3668	3702	3736	3770	3804	3837	3870	3903	3935	6	11	17	22	28
14	9.3968	4000	4032	4064	4095	4127	4158	4189	4220	4250	5	10	16	21	26
15	9.4281	4311	4341	4371	4400	4430	4459	4488	4517	4546	5	10	15	20	25
16	9.4575	4603	4632	4660	4688	4716	4744	4771	4799	4826	5	9	14	19	23
17	9.4853	4880	4907	4934	4961	4987	5014	5040	5066	5092	4	9	13	18	22

LOGARITHMIC TANGENTS (Continued)

Deg.	0′	6′	12′	18′	24′	30′	36′	42′	48′	54′	Mean Differences 1′	2′	3′	4′	5′
18	9.5118	5143	5169	5195	5220	5245	5270	5295	5320	5345	4	8	13	17	21
19	9.5370	5394	5419	5443	5467	5491	5516	5539	5563	5587	4	8	12	16	20
20	9.5611	5634	5658	5681	5704	5727	5750	5773	5796	5819	4	8	12	15	19
21	9.5842	5864	5887	5909	5932	5954	5976	5998	6020	6042	4	7	11	15	19
22	9.6064	6086	6108	6129	6151	6172	6194	6215	6236	6257	4	7	11	14	18
23	9.6279	6300	6321	6341	6362	6383	6404	6424	6445	6465	3	7	10	14	17
24	9.6486	6506	6527	6547	6567	6587	6607	6627	6647	6667	3	7	10	13	17
25	9.6687	6706	6726	6746	6765	6785	6804	6824	6843	6863	3	7	10	13	16
26	9.6882	6901	6920	6939	6958	6977	6996	7015	7034	7053	3	6	9	13	16
27	9.7072	7090	7109	7128	7146	7165	7183	7202	7220	7238	3	6	9	12	15
28	9.7257	7275	7293	7311	7330	7348	7366	7384	7402	7420	3	6	9	12	15
29	9.7438	7455	7473	7491	7509	7526	7544	7562	7579	7597	3	6	9	12	15
30	9.7614	7632	7649	7667	7684	7701	7719	7736	7753	7771	3	6	9	12	14
31	9.7788	7805	7822	7839	7856	7873	7890	7907	7924	7941	3	6	9	11	14
32	9.7958	7975	7992	8008	8025	8042	8059	8075	8092	8109	3	6	8	11	14
33	9.8125	8142	8158	8175	8191	8208	8224	8241	8257	8274	3	5	8	11	14
34	9.8290	8306	8323	8339	8355	8371	8388	8404	8420	8436	3	5	8	11	14
35	9.8452	8468	8484	8501	8517	8533	8549	8565	8581	8597	3	5	8	11	13

LOGARITHMIC TANGENTS (Continued)

Deg.	0′	6′	12′	18′	24′	30′	36′	42′	48′	54′	Mean Differences 1′	2′	3′	4′	5′
36	9.8613	8629	8644	8660	8676	8692	8708	8724	8740	8755	3	5	8	11	13
37	9.8771	8787	8803	8818	8834	8850	8865	8881	8897	8912	3	5	8	10	13
38	9.8928	8944	8959	8975	8990	9006	9022	9037	9053	9068	3	5	8	10	13
39	9.9084	9099	9115	9130	9146	9161	9176	9192	9207	9223	3	5	8	10	13
40	9.9238	9254	9269	9284	9300	9315	9330	9346	9361	9376	3	5	8	10	13
41	9.9392	9407	9422	9438	9453	9468	9483	9499	9514	9529	3	5	8	10	13
42	9.9544	9560	9575	9590	9605	9621	9636	9651	9666	9681	3	5	8	10	13
43	9.9697	9712	9727	9742	9757	9773	9788	9803	9818	9833	3	5	8	10	13
44	9.9848	9864	9870	9894	9909	9924	9939	9955	9970	9985	3	5	8	10	13
45	10.0000	0015	0030	0045	0061	0076	0091	0106	0121	0136	3	5	8	10	13
46	10.0152	0167	0182	0197	0212	0228	0243	0258	0273	0288	3	5	8	10	13
47	10.0303	0319	0334	0349	0364	0379	0395	0410	0425	0440	3	5	8	10	13
48	10.0456	0471	0486	0501	0517	0532	0547	0562	0578	0593	3	5	8	10	13
49	10.0608	0624	0639	0654	0670	0685	0700	0716	0731	0746	3	5	8	10	13
50	10.0762	0777	0793	0808	0824	0839	0854	0870	0885	0901	3	5	8	10	13
51	10.0916	0932	0947	0963	0978	0994	1010	1025	1041	1056	3	5	8	10	13
52	10.1072	1088	1103	1119	1135	1150	1166	1182	1197	1213	3	5	8	10	13
53	10.1229	1245	1260	1276	1292	1308	1324	1340	1356	1371	3	5	8	11	13

LOGARITHMIC TANGENTS (Continued)

Deg.	0′	6	12′	18′	24′	30′	36′	42′	48′	54′	Mean Differences. 1′	2′	3′	4′	5′
54	10.1387	1403	1419	1435	1451	1467	1483	1499	1516	1532	3	5	8	11	13
55	10.1548	1564	1580	1596	1612	1629	1645	1661	1677	1694	3	5	8	11	14
56	10.1710	1726	1743	1759	1776	1792	1809	1825	1842	1858	3	5	8	11	14
57	10.1875	1891	1908	1925	1941	1958	1975	1992	2008	2025	3	6	8	11	14
58	10.2042	2059	2076	2093	2110	2127	2144	2161	2178	2195	3	6	9	11	14
59	10.2212	2229	2247	2264	2281	2299	2316	2333	2351	2368	3	6	9	12	14
60	10.2386	2403	2421	2438	2456	2474	2491	2509	2527	2545	3	6	9	12	15
61	10.2562	2580	2598	2616	2634	2652	2670	2689	2707	2725	3	6	9	12	15
62	10.2743	2762	2780	2798	2817	2835	2854	2872	2891	2910	3	6	9	12	15
63	10.2928	2947	2966	2985	3004	3023	3042	3061	3080	3099	3	6	9	13	16
64	10.3118	3137	3157	3176	3196	3215	3235	3254	3274	3294	3	6	10	13	16
65	10.3313	3333	3353	3373	3393	3413	3433	3453	3473	3494	3	7	10	13	17
66	10.3514	3535	3555	3576	3596	3617	3638	3659	3679	3700	3	7	10	14	17
67	10.3721	3743	3764	3785	3806	3828	3849	3871	3892	3914	4	7	11	14	18
68	10.3936	3958	3980	4002	4024	4046	4068	4091	4113	4136	4	7	11	15	19
69	10.4158	4181	4204	4227	4250	4273	4296	4319	4342	4366	4	8	12	15	19
70	10.4389	4413	4437	4461	4484	4509	4533	4557	4581	4606	4	8	12	16	20
71	10.4630	4655	4680	4705	4730	4755	4780	4805	4831	4857	4	8	13	17	21

LOGARITHMIC TANGENTS—(Continued)

Deg.	0′	6′	12′	18′	24′	30′	36′	42′	48′	54′	Mean Differences 1′	2′	3′	4′	5′
72	10.4882	4908	4934	4960	4986	5013	5039	5066	5093	5120	4	9	13	18	22
73	10.5147	5174	5201	5229	5256	5284	5312	5430	5368	5397	5	9	14	19	23
74	10.5425	5454	5483	5512	5541	5570	5600	5629	5659	5689	5	10	15	20	25
75	10.5719	5750	5780	5811	5842	5873	5905	5936	5968	6000	5	10	16	21	26
76	10.6032	6065	6097	6130	6163	6196	6230	6264	6298	6332	6	11	17	22	28
77	10.6366	6401	6436	6471	6507	6542	6578	6615	6651	6688	6	12	18	24	30
78	10.6725	6763	6800	6838	6877	6915	6954	6994	7033	7073	6	13	19	26	32
79	10.7113	7154	7195	7236	7278	7320	7363	7406	7449	7493	7	14	21	28	35
80	10.7537	7581	7626	7672	7718	7764	7811	7858	7906	7954	8	16	23	31	39
81	10.8003	8052	8102	8152	8203	8255	8307	8360	8413	8467	9	17	26	35	43
82	10.8522	8577	8633	8690	8748	8806	8865	8924	8985	9046	10	20	29	39	49
83	10.9109	9172	9236	930[illegible]	9367	9433	9501	9570	9640	9711	11	22	34	45	56
84	10.9784	9857	9932	$\overline{0008}$	$\overline{0085}$	$\overline{0164}$	$\overline{0244}$	$\overline{0326}$	$\overline{0409}$	$\overline{0494}$	13	26	40	53	66
85	11.0580	0669	0759	0850	0944	1040	1138	1238	1341	1446	16	32	48	64	81
86	11.1554	1664	1777	1893	2012	2135	2261	2391	2525	2663	*Mean* differences no longer sufficiently accurate.				
87	11.2806	2954	3106	3264	3429	3599	3777	3962	4155	4357					
88	11.4569	4792	5027	5275	5539	5189	6119	6441	6789	7167					
89	11.7581	8038	8550	9130	9800	$\overline{0591}$	$\overline{1561}$	$\overline{2810}$	$\overline{4571}$	$\overline{7581}$					

CHORDS OF CIRCLES.

No. of Spaces	Multiply Dia. by	No. of Spaces	Multiply Dia. by	No. of Spaces	Multiply Dia. by	No. of Spaces	Multiply Dia. by	No. of Spaces	Multiply Dia. by
3	.8660	23	.1362	43	.0730	63	.0499	83	.0378
4	.7071	24	.1305	44	.0713	64	.9491	84	.0374
5	.5878	25	.1253	45	.0698	65	.0483	85	.0370
6	.5000	26	.1205	46	.0682	66	.0476	86	.0365
7	.4339	27	.1161	47	.0668	67	.0469	87	.0361
8	.3827	28	.1120	48	.0654	68	.0462	88	.0357
9	.3420	29	.1081	49	.0641	69	.0455	89	.0353
10	.3090	30	.1045	50	.0628	70	.0449	90	.0349
11	.2817	31	.1012	51	.0616	71	.0442	91	.0345
12	.2588	32	.0980	52	.0604	72	.0436	92	.0341
13	.2393	33	.0951	53	.0592	73	.0430	93	.0338
14	.2225	34	.0923	54	.0581	74	.0424	94	.0334
15	.2079	35	.0896	55	.0571	75	.0419	95	.0331
16	.1951	36	.0872	56	.0561	76	.0413	96	.0327
17	.1838	37	.0848	57	.0551	77	.0408	97	.0324
18	.1736	38	.0826	58	.0541	78	.0403	98	.0321
19	.1646	39	.0805	59	.0532	79	.0398	99	.0317
20	.1564	40	.0785	60	.0523	80	.0393	100	.0314
21	.1490	41	.0765	61	.0515	81	.0388		
22	.1423	42	.0747	62	.0507	82	.0383		

POWERS AND ROOTS

No.	Squares	Cubes	Square Roots	Cube Roots
1	1	1	1.000	1.000
2	4	8	1.414	1.260
3	9	27	1.732	1.442
4	16	64	2.000	1.587
5	25	125	2.236	1.710
6	36	216	2.449	1.817
7	49	343	2.646	1.913
8	64	512	2.828	2.000
9	81	729	3.000	2.080
10	100	1 000	3.162	2.154
11	121	1 331	3.317	2.224
12	144	1 728	3.464	2.289
13	169	2 197	3.606	2.351
14	196	2 744	3.742	2.410
15	225	3 375	3.873	2.466
16	256	4 096	4.000	2.520
17	289	4 913	4.123	2.571
18	324	5 832	4.243	2.621
19	361	6 859	4.359	2.668
20	400	8 000	4.472	2.714
21	441	9 261	4.583	2.759
22	484	10 648	4.690	2.802
23	529	12 167	4.796	2.844
24	576	13 824	4.899	2.884
25	625	15 625	5.000	2.924
26	676	17 576	5.099	2.962
27	729	19 683	5.196	3.000
28	784	21 952	5.292	3.037
29	841	24 389	5.385	3.072
30	900	27 000	5.477	3.107
31	961	29 791	5.568	3.141
32	1 024	32 768	5.657	3.175
33	1 089	35 937	5.745	3.208
34	1 156	39 304	5.831	3.240
35	1 225	42 875	5.916	3.271

POWERS AND ROOTS *(Continued)*

No.	Squares	Cubes	Square Roots	Cube Roots
36	1 296	46 656	6.000	3.302
37	1 369	50 653	6.083	3.332
38	1 444	54 872	6.164	3.362
39	1 521	59 319	6.245	3.391
40	1 600	64 000	6.325	3.420
41	1 681	68 921	6.403	3.448
42	1 764	74 088	6.481	3.476
43	1 849	79 507	6.557	3.503
44	1 936	85 184	6.633	3.530
45	2 025	91 125	6.708	3.557
46	2 116	97 336	6.782	3.583
47	2 209	103 823	6.856	3.609
48	2 304	110 592	6.928	3.634
49	2 401	117 649	7.000	3.659
50	2 500	125 000	7.071	3.684
51	2 601	132 651	7.141	3.708
52	2 704	140 608	7.211	3.733
53	2 809	148 877	7.280	3.756
54	2 916	157 464	7.348	3.780
55	3 025	166 375	7.416	3.803
56	3 136	175 616	7.483	3.826
57	3 249	185 193	7.550	3.849
58	3 364	195 112	7.616	3.871
59	3 481	205 379	7.681	3.893
60	3 600	216 000	7.746	3.915
61	3 721	226 981	7.810	3.936
62	3 844	238 328	7.874	3.958
63	3 969	250 047	7.937	3.979
64	4 096	262 144	8.000	4.000
65	4 225	274 625	8.062	4.021
66	4 356	287 496	8.124	4.041
67	4 489	300 763	8.185	4.062
68	4 624	314 432	8.246	4.082
69	4 761	328 509	8.307	4.102

POWERS AND ROOTS (*Continued*)

No.	Squares	Cubes	Square Roots	Cube Roots
70	4 900	343 000	8.367	4.121
71	5 041	357 911	8.426	4.141
72	5 184	373 248	8.485	4.160
73	5 329	389 017	8.544	4.179
74	5 476	405 224	8.602	4.198
75	5 625	421 875	8.660	4.217
76	5 776	438 976	8.718	4.236
77	5 929	456 533	8.775	4.254
78	6 084	474 552	8.832	4.273
79	6 241	493 039	8.888	4.291
80	6 400	512 000	8.944	4.309
81	6 561	531 441	9.000	4.327
82	6 724	551 368	9.055	4.344
83	6 889	571 787	9.110	4.362
84	7 056	592 704	9.165	4.380
85	7 225	614 125	9.220	4.397
86	7 396	636 056	9.274	4.414
87	7 569	658 503	9.327	4.431
88	7 744	681 472	9.381	4.448
89	7 921	704 969	9.434	4.465
90	8 100	729 000	9.487	4.481
91	8 281	753 571	9.539	4.498
92	8 464	778 688	9.592	4.514
93	8 649	804 357	9.644	4.531
94	8 836	830 584	9.695	4.547
95	9 025	857 375	9.747	4.563
96	9 216	884 736	9.798	4.579
97	9 409	912 673	9.849	4.595
98	9 604	941 192	9.899	4.610
99	9 801	970 299	9.950	4.626
100	10 000	1 000 000	10.000	4.642

PRIME NUMBERS AND FACTORS

From 1 to 100

1		26	2×13	51	3×17	76	$2^2 \times 19$
2		27	3^3	52	$2^2 \times 13$	77	7×11
3		28	$2^2 \times 7$	53		78	$2 \times 3 \times 13$
4	2^2	29		54	2×3^3	79	
5		30	$2 \times 3 \times 5$	55	5×11	80	$2^4 \times 5$
6	2×3	31		56	$2^3 \times 7$	81	3^4
7		32	2^5	57	3×19	82	2×41
8	2^3	33	3×11	58	2×29	83	
9	3^2	34	2×17	59		84	$2^2 \times 3 \times 7$
10	2×5	35	5×7	60	$2^2 \times 3 \times 5$	85	5×17
11		36	$2^2 \times 3^2$	61		86	2×43
12	$2^2 \times 3$	37		62	2×31	87	3×29
13		38	2×19	63	$3^2 \times 7$	88	$2^3 \times 11$
14	2×7	39	3×13	64	2^6	89	
15	3×5	40	$2^3 \times 5$	65	5×13	90	$2 \times 3^2 \times 5$
16	2^4	41		66	$2 \times 3 \times 11$	91	7×13
17		42	$2 \times 3 \times 7$	67		92	$2^2 \times 23$
18	2×3^2	43		68	$2^2 \times 17$	93	3×31
19		44	$2^2 \times 11$	69	3×23	94	2×47
20	$2^2 \times 5$	45	$3^2 \times 5$	70	$2 \times 5 \times 7$	95	5×19
21	3×7	46	2×23	71		96	$2^5 \times 3$
22	2×11	47		72	$2^3 \times 3^2$	97	
23		48	$2^4 \times 3$	73		98	2×7^2
24	$2^3 \times 3$	49	7^2	74	2×37	99	$3^2 \times 11$
25	5^2	50	2×5^2	75	3×5^2	100	$2^2 \times 5^2$

POWERS AND ROOTS OF π and g

n	$\frac{1}{n}$	n^2	n^3	$\sqrt{n}$	$\frac{1}{\sqrt{n}}$	$\sqrt[3]{n}$	$\frac{1}{\sqrt[3]{n}}$
$\pi=3.142$	0.318	9.870	31.006	1.772	0.564	1.465	0.683
$2\pi=6.283$	0.159	39.478	248.050	2.507	0.399	1.845	0.542
$\frac{\pi}{2}=1.571$	0.637	2.467	3.878	1.253	0.798	1.162	0.860
$\frac{\pi}{3}=1.047$	0.955	1.097	1.148	1.023	0.977	1.016	0.985
$\frac{4}{3}\pi=4.189$	0.239	17.546	73.496	2.047	0.489	1.612	0.622
$\frac{\pi}{4}=0.785$	1.274	0.617	0.484	0.886	1.128	0.923	1.084
$\frac{\pi}{6}=0.524$	1.910	0.274	0.144	0.724	1.382	0.806	1.241
$\pi^2=9.870$	0.101	97.409	961.390	3.142	0.318	2.145	0.466
$\pi^3=31.066$	0.032	961.390	29.809.910	5.568	1.796	3.142	0.318
$\frac{\pi}{32}=0.098$	10.186	0.0095	0.001	0.313	3.192	0.461	2.168
g 32.2	0.031	1036.84	33,386.24	5.674	0.176	3.181	0.314
2g 64.4	0.015	4147.36	267 090	8.025	0.125	4.007	0.249

LOGARITHMS

	0	1	2	3	4	5	6	7	8	9	1	2	3	4	5	6	7	8	9
10	0000	0043	0086	0128	0170	0212	0253	0294	0334	0374	4	8	12	17	21	25	29	33	37
11	0414	0453	0492	0531	0569	0607	0645	0682	0719	0755	4	8	11	15	19	23	26	30	34
12	0792	0828	0864	0899	0934	0969	1004	1038	1072	1106	3	7	10	14	17	21	24	28	31
13	1139	1173	1206	1239	1271	1303	1335	1367	1399	1430	3	6	10	13	16	19	23	26	29
14	1461	1492	1523	1553	1584	1614	1644	1673	1703	1732	3	6	9	12	15	18	21	24	27
15	1761	1790	1818	1847	1875	1903	1931	1959	1987	2014	3	6	8	11	14	17	20	22	25
16	2041	2068	2095	2122	2148	2175	2201	2227	2253	2279	3	5	8	11	13	16	18	21	24
17	2304	2330	2355	2380	2405	2430	2455	2480	2504	2529	2	5	7	10	12	15	17	20	22
18	2553	2577	2601	2625	2648	2672	2695	2718	2742	2765	2	5	7	9	12	14	16	19	21
19	2788	2810	2833	2856	2878	2900	2923	2945	2967	2989	2	4	7	9	11	13	16	18	20
20	3010	3032	3054	3075	3096	3118	3139	3160	3181	3201	2	4	6	8	11	13	15	17	19
21	3222	3243	3263	3284	3304	3324	3345	3365	3385	3404	2	4	6	8	10	12	14	16	18
22	3424	3444	3464	3483	3502	3522	3541	3560	3579	3598	2	4	6	8	10	12	14	15	17
23	3617	3636	3655	3674	3692	3711	3729	3747	3766	3784	2	4	6	7	9	11	13	15	17
24	3802	3820	3838	3856	3874	3892	3909	3927	3945	3962	2	4	5	7	9	11	12	14	16
25	3979	3997	4014	4031	4048	4065	4082	4099	4116	4133	2	3	5	7	9	10	12	14	15
26	4150	4166	4183	4200	4216	4232	4249	4265	4281	4298	2	3	5	7	8	10	11	13	15
27	4314	4330	4346	4362	4378	4393	4409	4425	4440	4456	2	3	5	6	8	9	11	13	14
28	4472	4487	4502	4518	4533	4548	4564	4579	4594	4609	2	3	5	6	8	9	11	12	14
29	4624	4639	4654	4669	4683	4698	4713	4728	4742	4757	1	3	4	6	7	9	10	12	13

LOGARITHMS—(*continued*)

	0	1	2	3	4	5	6	7	8	9	1	2	3	4	5	6	7	8	9
30	4771	4786	4800	4814	4829	4843	4857	4871	4886	4900	1	3	4	6	7	9	10	11	13
31	4914	4928	4942	4955	4969	4983	4997	5011	5024	5038	1	3	4	6	7	8	10	11	12
32	5051	5065	5079	5092	5105	5119	5132	5145	5159	5172	1	3	4	5	7	8	9	11	12
33	5185	5198	5211	5224	5237	5250	5263	5276	5289	5302	1	3	4	5	6	8	9	10	12
34	5315	5328	5340	5353	5366	5378	5391	5403	5416	5428	1	3	4	5	6	8	9	10	11
35	5441	5453	5465	5478	5490	5502	5514	5527	5539	5551	1	2	4	5	6	7	9	10	11
36	5563	5575	5587	5599	5611	5623	5635	5647	5658	5670	1	2	4	5	6	7	8	10	11
37	5682	5694	5705	5717	5729	5740	5752	5763	5775	5786	1	2	3	5	6	7	8	9	10
38	5798	5809	5821	5832	5843	5855	5866	5877	5888	5899	1	2	3	5	6	7	8	9	10
39	5911	5922	5933	5944	5955	5966	5977	5988	5999	6010	1	2	3	4	5	7	8	9	10
40	6021	6031	6042	6053	6064	6075	6085	6096	6107	6117	1	2	3	4	5	6	8	9	10
41	6128	6138	6149	6160	6170	6180	6191	6201	6212	6222	1	2	3	4	5	6	7	8	9
42	6232	6243	6253	6263	6274	6284	6294	6304	6314	6325	1	2	3	4	5	6	7	8	9
43	6335	6345	6355	6365	6375	6385	6395	6405	6415	6425	1	2	3	4	5	6	7	8	9
44	6435	6444	6454	6464	6474	6484	6493	6503	6513	6522	1	2	3	4	5	6	7	8	9
45	6532	6542	6551	6561	6571	6580	6590	6599	6609	6618	1	2	3	4	5	6	7	8	9
46	6628	6637	6646	6656	6665	6675	6684	6693	6702	6712	1	2	3	4	5	6	7	7	8
47	6721	6730	6739	6749	6758	6767	6776	6785	6794	6803	1	2	3	4	5	5	6	7	8
48	6812	6821	6830	6839	6848	6857	6866	6875	6884	6893	1	2	3	4	4	5	6	7	8
49	6902	6911	6920	6928	6937	6946	6955	6964	6972	6981	1	2	3	4	4	5	6	7	8
50	6990	6998	7007	7016	7024	7033	7042	7050	7059	7067	1	2	3	3	4	5	6	7	8
51	7076	7084	7093	7101	7110	7118	7126	7135	7143	7152	1	2	3	3	4	5	6	7	8
52	7160	7168	7177	7185	7193	7202	7210	7218	7226	7235	1	2	2	3	4	5	6	7	7
53	7243	7251	7259	7267	7275	7284	7292	7300	7308	7316	1	2	2	3	4	5	6	6	7
54	7324	7332	7340	7348	7356	7364	7372	7380	7388	7396	1	2	2	3	4	5	6	6	7

LOGARITHMS—*(continued)*

	0	1	2	3	4	5	6	7	8	9	1	2	3	4	5	6	7	8	9
55	7404	7412	7419	7427	7435	7443	7451	7459	7466	7474	1	2	2	3	4	5	5	6	7
56	7482	7490	7497	7505	7513	7520	7528	7536	7543	7551	1	2	2	3	4	5	5	6	7
57	7559	7566	7574	7582	7589	7597	7604	7612	7619	7627	1	2	2	3	4	5	5	6	7
58	7634	7642	7649	7657	7664	7672	7679	7686	7694	7701	1	1	2	3	4	4	5	6	7
59	7709	7716	7723	7731	7738	7745	7752	7760	7767	7774	1	1	2	3	4	4	5	6	7
60	7782	7789	7796	7803	7810	7818	7825	7832	7839	7846	1	1	2	3	4	4	5	6	6
61	7853	7860	7868	7875	7882	7889	7896	7903	7910	7917	1	1	2	3	4	4	5	6	6
62	7924	7931	7938	7945	7952	7959	7966	7973	7980	7987	1	1	2	3	3	4	5	6	6
63	7993	8000	8007	8014	8021	8028	8035	8041	8048	8055	1	1	2	3	3	4	5	5	6
64	8062	8069	8075	8082	8089	8096	8102	8109	8116	8122	1	1	2	3	3	4	5	5	6
65	8129	8136	8142	8149	8156	8162	8169	8176	8182	8189	1	1	2	3	3	4	5	5	6
66	8195	8202	8209	8215	8222	8228	8235	8241	8248	8254	1	1	2	3	3	4	5	5	6
67	8261	8267	8274	8280	8287	8293	8299	8306	8312	8319	1	1	2	3	3	4	5	5	6
68	8325	8331	8338	8344	8351	8357	8363	8370	8376	8382	1	1	2	3	3	4	4	5	6
69	8388	8395	8401	8407	8414	8420	8426	8432	8439	8445	1	1	2	2	3	4	4	5	6
70	8451	8457	8463	8470	8476	8482	8488	8494	8500	8506	1	1	2	2	3	4	4	5	6
71	8513	8519	8525	8531	8537	8543	8549	8555	8561	8567	1	1	2	2	3	4	4	5	5
72	8573	8579	8585	8591	8597	8603	8609	8615	8621	8627	1	1	2	2	3	4	4	5	5
73	8633	8639	8645	8651	8657	8663	8669	8675	8681	8686	1	1	2	2	3	4	4	5	5
74	8692	8698	8704	8710	8716	8722	8727	8733	8739	8745	1	1	2	2	3	4	4	5	5

LOGARITHMS—(*continued*)

	0	1	2	3	4	5	6	7	8	9	1	2	3	4	5	6	7	8	9
75	8751	8756	8762	8768	8774	8779	8785	8791	8797	8802	1	1	2	2	3	3	4	5	5
76	8808	8814	8820	8825	8831	8837	8842	8848	8854	8859	1	1	2	2	3	3	4	5	5
77	8865	8871	8876	8882	8887	8893	8899	8904	8910	8915	1	1	2	2	3	3	4	4	5
78	8921	8927	8932	8938	8943	8949	8954	8960	8965	8971	1	1	2	2	3	3	4	4	5
79	8976	8982	8987	8993	8998	9004	9009	9015	9020	9025	1	1	2	2	3	3	4	4	5
80	9031	9036	9042	9047	9053	9058	9063	9069	9074	9079	1	1	2	2	3	3	4	4	5
81	9085	9090	9096	9101	9106	9112	9117	9122	9128	9133	1	1	2	2	3	3	4	4	5
82	9138	9143	9149	9154	9159	9165	9170	9175	9180	9186	1	1	2	2	3	3	4	4	5
83	9191	9196	9201	9206	9212	9217	9222	9227	9232	9238	1	1	2	2	3	3	4	4	5
84	9243	9248	9253	9258	9263	9269	9274	9279	9284	9289	1	1	2	2	3	3	4	4	5
85	9294	9299	9304	9309	9315	9320	9325	9330	9335	9340	1	1	2	2	3	3	4	4	5
86	9345	9350	9355	9360	9365	9370	9375	9380	9385	9390	1	1	2	2	3	3	4	4	5
87	9395	9400	9405	9410	9415	9420	9425	9430	9435	9440	0	1	1	2	2	3	3	4	4
88	9445	9450	9455	9460	9465	9469	9474	9479	9484	9489	0	1	1	2	2	3	3	4	4
89	9494	9499	9504	9509	9513	9518	9523	9528	9533	9538	0	1	1	2	2	3	3	4	4
90	9542	9547	9552	9557	9562	9566	9571	9576	9581	9586	0	1	1	2	2	3	3	4	4
91	9590	9595	9600	9605	9609	9614	9619	9624	9628	9633	0	1	1	2	2	3	3	4	4
92	9638	9643	9647	9652	9657	9661	9666	9671	9675	9680	0	1	1	2	2	3	3	4	4
93	9685	9689	9694	9699	9703	9708	9713	9717	9722	9727	0	1	1	2	2	3	3	4	4
94	9731	9736	9741	9745	9750	9754	9759	9763	9768	9773	0	1	1	2	2	3	3	4	4
95	9777	9782	9786	9791	9795	9800	9805	9809	9814	9818	0	1	1	2	2	3	3	4	4
96	9823	9827	9832	9836	9841	9845	9850	9854	9859	9863	0	1	1	2	2	3	3	4	4
97	9868	9872	9877	9881	9886	9890	9894	9899	9903	9908	0	1	1	2	2	3	3	4	4
98	9912	9917	9921	9926	9930	9934	9939	9943	9948	9952	0	1	1	2	2	3	3	4	4
99	9956	9961	9965	9969	9974	9978	9983	9987	9991	9996	0	1	1	2	2	3	3	3	4

ANTILOGARITHMS

	0	1	2	3	4	5	6	7	8	9	1	2	3	4	5	6	7	8	9
.00	1000	1002	1005	1007	1009	1012	1014	1016	1019	1021	0	0	1	1	1	1	2	2	2
.01	1023	1026	1028	1030	1033	1035	1038	1040	1042	1045	0	0	1	1	1	1	2	2	2
.02	1047	1050	1052	1054	1057	1059	1062	1064	1067	1069	0	0	1	1	1	1	2	2	2
.03	1072	1074	1076	1079	1081	1084	1086	1089	1091	1094	0	0	1	1	1	1	2	2	2
.04	1096	1099	1102	1104	1107	1109	1112	1114	1117	1119	0	1	1	1	1	2	2	2	2
.05	1122	1125	1127	1130	1132	1135	1138	1140	1143	1146	0	1	1	1	1	2	2	2	2
.06	1148	1151	1153	1156	1159	1161	1164	1167	1169	1172	0	1	1	1	1	2	2	2	2
.07	1175	1178	1180	1183	1186	1189	1191	1194	1197	1199	0	1	1	1	1	2	2	2	2
.08	1202	1205	1208	1211	1213	1216	1219	1222	1225	1227	0	1	1	1	1	2	2	2	3
.09	1230	1233	1236	1239	1242	1245	1247	1250	1253	1256	0	1	1	1	1	2	2	2	3
.10	1259	1262	1265	1268	1271	1274	1276	1279	1282	1285	0	1	1	1	1	2	2	2	3
.11	1288	1291	1294	1297	1300	1303	1306	1309	1312	1315	0	1	1	1	2	2	2	2	3
.12	1318	1321	1324	1327	1330	1334	1337	1340	1343	1346	0	1	1	1	2	2	2	2	3
.13	1349	1352	1355	1358	1361	1365	1368	1371	1374	1377	0	1	1	1	2	2	2	3	3
.14	1380	1384	1387	1390	1393	1396	1400	1403	1406	1409	0	1	1	1	2	2	2	3	3
.15	1413	1416	1419	1422	1426	1429	1432	1435	1439	1442	0	1	1	1	2	2	2	3	3
.16	1445	1449	1452	1455	1459	1462	1466	1469	1472	1476	0	1	1	1	2	2	2	3	3
.17	1479	1483	1486	1489	1493	1496	1500	1503	1507	1510	0	1	1	1	2	2	2	3	3
.18	1514	1517	1521	1524	1528	1531	1535	1538	1542	1545	0	1	1	1	2	2	2	3	3
.19	1549	1552	1556	1560	1563	1567	1570	1574	1578	1581	0	1	1	1	2	2	3	3	3
.20	1585	1589	1592	1596	1600	1603	1607	1611	1614	1618	0	1	1	1	2	2	3	3	3
.21	1622	1626	1629	1633	1637	1641	1644	1648	1652	1656	0	1	1	2	2	2	3	3	3
.22	1660	1663	1667	1671	1675	1679	1683	1687	1690	1694	0	1	1	2	2	2	3	3	3
.23	1698	1702	1706	1710	1714	1718	1722	1726	1730	1734	0	1	1	2	2	2	3	3	4

ANTILOGARITHMS—(*continued*).

	0	1	2	3	4	5	6	7	8	9	1	2	3	4	5	6	7	8	9
.24	1738	1742	1746	1750	1754	1758	1762	1766	1770	1774	0	1	1	2	2	2	3	3	4
.25	1778	1782	1786	1791	1795	1799	1803	1807	1811	1816	0	1	1	2	2	2	3	3	4
.26	1820	1824	1828	1832	1837	1841	1845	1849	1854	1858	0	1	1	2	2	3	3	3	4
.27	1862	1866	1871	1875	1879	1884	1888	1892	1897	1901	0	1	1	2	2	3	3	3	4
.28	1905	1910	1914	1919	1923	1928	1932	1936	1941	1945	0	1	1	2	2	3	3	4	4
.29	1950	1954	1959	1963	1968	1972	1977	1982	1986	1991	0	1	1	2	2	3	3	4	4
.30	1995	2000	2004	2009	2014	2018	2023	2028	2032	2037	0	1	1	2	2	3	3	4	4
.31	2042	2046	2051	2056	2061	2065	2070	2075	2080	2084	0	1	1	2	2	3	3	4	4
.32	2089	2094	2099	2104	2109	2113	2118	2123	2128	2133	0	1	1	2	2	3	3	4	4
.33	2138	2143	2148	2153	2158	2163	2168	2173	2178	2183	0	1	1	2	2	3	3	4	4
.34	2188	2193	2198	2203	2208	2213	2218	2223	2228	2234	1	1	2	2	3	3	4	4	5
.35	2239	2244	2249	2254	2259	2265	2270	2275	2280	2286	1	1	2	2	3	3	4	4	5
.36	2291	2296	2301	2307	2312	2317	2323	2328	2333	2339	1	1	2	2	3	3	4	4	5
.37	2344	2350	2355	2360	2366	2371	2377	2382	2388	2393	1	1	2	2	3	3	4	4	5
.38	2399	2404	2410	2415	2421	2427	2432	2438	2443	2449	1	1	2	2	3	3	4	4	5
.39	2455	2460	2466	2472	2477	2483	2489	2495	2500	2506	1	1	2	2	3	3	4	5	5
.40	2512	2518	2523	2529	2535	2541	2547	2553	2559	2564	1	1	2	2	3	4	4	5	5
.41	2570	2576	2582	2588	2594	2600	2 06	2612	2618	2624	1	1	2	2	3	4	4	5	5
.42	2630	2636	2642	2649	2655	2661	2667	2673	2679	2685	1	1	2	2	3	4	4	5	6
.43	2692	2698	2704	2710	2716	2723	2729	2735	2742	2748	1	1	2	3	3	4	4	5	6
.44	2754	2761	2767	2773	2780	2786	2793	2799	2805	2812	1	1	2	3	3	4	4	5	6
.45	2818	2825	2831	2838	2844	2851	2858	2864	2871	2877	1	1	2	3	3	4	5	5	6
.46	2884	2891	2897	2904	2911	2917	2924	2931	2938	2944	1	1	2	3	3	4	5	5	6
.47	2951	2958	2965	2972	2979	2985	2992	2999	3006	3013	1	1	2	3	3	4	5	5	6
.48	3020	3027	3034	3041	3048	3055	3062	3069	3076	3083	1	1	2	3	4	4	5	6	6
.49	3090	3097	3105	3112	3119	3126	3133	3141	3148	3155	1	1	2	3	4	4	5	6	6

ANTILOGARITHMS—(*continued*).

	0	1	2	3	4	5	6	7	8	9	1	2	3	4	5	6	7	8	9
.50	3162	3170	3177	3184	3192	3199	3206	3214	3221	3228	1	1	2	3	4	4	5	6	7
.51	3236	3243	3251	3258	3266	3273	3281	3289	3296	3304	1	2	2	3	4	5	5	6	7
.52	3311	3319	3327	3334	3342	3350	3357	3365	3373	3381	1	2	2	3	4	5	5	6	7
.53	3388	3396	3404	3412	3420	3428	3436	3443	3451	3459	1	2	2	3	4	5	6	6	7
.54	3467	3475	3483	3491	3499	3508	3516	3524	3532	3540	1	2	2	3	4	5	6	6	7
.55	3548	3556	3565	3573	3581	3589	3597	3606	3614	3622	1	2	2	3	4	5	6	7	7
.56	3631	3639	3648	3656	3664	3673	3681	3690	3698	3707	1	2	3	3	4	5	6	7	8
.57	3715	3724	3733	3741	3750	3758	3767	3776	3784	3793	1	2	3	3	4	5	6	7	8
.58	3802	3811	3819	3828	3837	3846	3855	3864	3873	3882	1	2	3	4	4	5	6	7	8
.59	3890	3899	3908	3917	3926	3936	3945	3954	3963	3972	1	2	3	4	5	5	6	7	8
.60	3981	3990	3999	4009	4018	4027	4036	4046	4055	4064	1	2	3	4	5	6	6	7	8
.61	4074	4083	4093	4102	4111	4121	4130	4140	4150	4159	1	2	3	4	5	6	7	8	9
.62	4169	4178	4188	4198	4207	4217	4227	4236	4246	4256	1	2	3	4	5	6	7	8	9
.63	4266	4276	4285	4295	4305	4315	4325	4335	4345	4355	1	2	3	4	5	6	7	8	9
.64	4365	4375	4385	4395	4406	4416	4426	4436	4446	4457	1	2	3	4	5	6	7	8	9
.65	4467	4477	4487	4498	4508	4519	4529	4539	4550	4560	1	2	3	4	5	6	7	8	9
.66	4571	4581	4592	4603	4613	4624	4634	4645	4656	4667	1	2	3	4	5	6	7	9	10
.67	4677	4688	4699	4710	4721	4732	4742	4753	4764	4775	1	2	3	4	5	7	8	9	10
.68	4786	4797	4808	4819	4831	4842	4853	4864	4875	4887	1	2	3	4	6	7	8	9	10
.69	4898	4909	4920	4932	4943	4955	4966	4977	4989	5000	1	2	3	5	6	7	8	9	10
.70	5012	5023	5035	5047	5058	5070	5082	5093	5105	5117	1	2	4	5	6	7	8	9	11
.71	5129	5140	5152	5164	5176	5188	5200	5212	5224	5236	1	2	4	5	6	7	8	10	11
.72	5248	5260	5272	5284	5297	5309	5321	5333	5346	5358	1	2	4	5	6	7	9	10	11
.73	5370	5383	5395	5408	5420	5433	5445	5458	5470	5483	1	3	4	5	6	8	9	10	11
.74	5495	5508	5521	5534	5546	5559	5572	5585	5598	5610	1	3	4	5	6	8	9	10	12

ANTILOGARITHMS—(*continued*).

	0	1	2	3	4	5	6	7	8	9	1	2	3	4	5	6	7	8	9
.75	5623	5636	5649	5662	5675	5689	5702	5715	5728	5741	1	3	4	5	7	8	9	10	12
.76	5754	5768	5781	5794	5808	5821	5834	5848	5861	5875	1	3	4	5	7	8	9	11	12
.77	5888	5902	5916	5929	5943	5957	5970	5984	5998	6012	1	3	4	5	7	8	10	11	12
.78	6026	6039	6053	6067	6081	6095	6109	6124	6138	6152	1	3	4	6	7	8	10	11	13
.79	6166	6180	6194	6209	6223	6237	6252	6266	6281	6295	1	3	4	6	7	9	10	11	13
.80	6310	6324	6339	6353	6368	6383	6397	6412	6427	6442	1	3	4	6	7	9	10	12	13
.81	6457	6471	6486	6501	6516	6531	6546	6561	6577	6592	2	3	5	6	8	9	11	12	14
.82	6607	6622	6637	6653	6668	6683	6699	6714	6730	6745	2	3	5	6	8	9	11	12	14
.83	6761	6776	6792	6808	6823	6839	6855	6871	6887	6902	2	3	5	6	8	9	11	13	14
.84	6918	6934	6950	6966	6982	6998	7015	7031	7047	7063	2	3	5	6	8	10	11	13	15
.85	7079	7096	7112	7129	7145	7161	7178	7194	7211	7228	2	3	5	7	8	10	12	13	15
.86	7244	7261	7278	7295	7311	7328	7345	7362	7379	7396	2	3	5	7	8	10	12	13	15
.87	7413	7430	7447	7464	7482	7499	7516	7534	7551	7568	2	4	5	7	9	10	12	14	16
.88	7586	7603	7621	7638	7656	7674	7691	7709	7727	7745	2	4	5	7	9	11	12	14	16
.89	7762	7780	7798	7816	7834	7852	7870	7889	7907	7925	2	4	5	7	9	11	13	14	16
.90	7943	7962	7980	7998	8017	8035	8054	8072	8091	8110	2	4	6	7	9	11	13	15	17
.91	8128	8147	8166	8185	8204	8222	8241	8260	8279	8299	2	4	6	8	9	11	13	15	17
.92	8318	8337	8356	8375	8395	8414	8433	8453	8472	8492	2	4	6	8	10	12	14	15	17
.93	8511	8531	8551	8570	8590	8610	8630	8650	8670	8690	2	4	6	8	10	12	14	16	18
.94	8710	8730	8750	8770	8790	8810	8831	8851	8872	8892	2	4	6	8	10	12	14	16	18
.95	8913	8933	8954	8974	8995	9016	9036	9057	9078	9099	2	4	6	8	10	12	15	17	19
.96	9120	9141	9162	9183	9204	9226	9247	9268	9290	9311	2	4	6	8	11	13	15	17	19
.97	9333	9354	9376	9397	9419	9441	9462	9484	9506	9528	2	4	7	9	11	13	15	17	20
.98	9550	9572	9594	9616	9638	9661	9683	9705	9727	9750	2	4	7	9	11	13	16	18	20
.99	9772	9795	9817	9840	9863	9886	9908	9931	9954	9977	2	5	7	9	11	14	16	18	20

TABLE OF DECIMAL EQUIVALENTS

Fraction	Decimal	Fraction	Decimal
1/64	·015625	33/64	·515625
1/32	·03125	17/32	·53125
3/64	·046875	35/64	·546875
1/16	·0625	9/16	·5625
5/64	·078125	37/64	·578125
3/32	·09375	19/32	·59375
7/64	·109375	39/64	·609375
1/8	·1250	5/8	·6250
9/64	·140625	41/64	·640625
5/32	.15625	21/32	·65625
11/64	·171875	43/64	·671875
3/16	·1875	11/16	·6875
13/64	·203125	45/64	·703125
7/32	·21875	23/32	·71875
15/64	·234375	47/64	·734375
1/4	·2500	3/4	·7500
17/64	·265625	49/64	·765625
9/32	·28125	25/32	·78125
19/64	·296875	51/64	·796875
5/16	·3125	13/16	·8125
21/64	·328125	53/64	·828125
11/32	·34375	27/32	·84375
23/64	·359375	55/64	·859375
3/8	·375	7/8	·8750
25/64	·390625	57/64	·890625
13/32	·40625	29/32	·90625
27/64	·421875	59/64	·921875
7/16	·4375	15/16	·9375
29/64	·453125	61/64	·953125
15/32	·46875	31/32	·96875
31/64	·484375	63/64	·984375
1/2	·5000	1	1·0000

INDEX

A

B

C

D

E

F

G

H

I

J

K

L

M

N

S

T

X

Y

Z

Index

Index